Swept-Source Optical Coherence Tomography

A Color Atlas

Second Edition

Swept-Source Optical Coherence Tomography

A Color Atlas

Second Edition

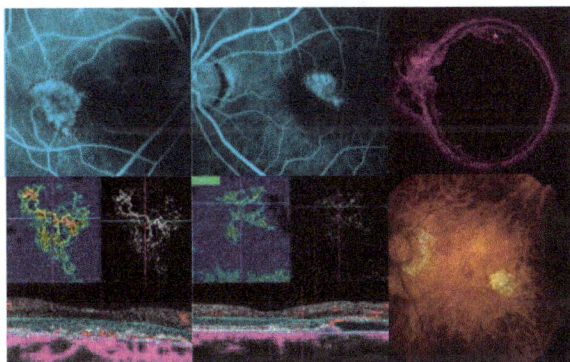

Kelvin Y.C. Teo
Wong Chee Wai
Andrew S.H. Tsai
Daniel S.W. Ting
Dan Milea
Singapore National Eye Centre, Singapore

Edited by

Lee Shu Yen and Gemmy C.M. Cheung
Singapore National Eye Centre, Singapore

Foreword by

Wong Tien Yin
Singapore National Eye Centre, Singapore

Singapore National Eye Centre

World Scientific

Published by

World Scientific Publishing Co. Pte. Ltd.

5 Toh Tuck Link, Singapore 596224

USA office: 27 Warren Street, Suite 401-402, Hackensack, NJ 07601

UK office: 57 Shelton Street, Covent Garden, London WC2H 9HE

Library of Congress Cataloging-in-Publication Data
Names: Teo, Kelvin Y. C., author. | Wong, Chee Wai, author. |
    Tsai, Andrew S. H., author. | Ting, Daniel S. W., author. | Milea, Dan, author.
Title: Swept-source optical coherence tomography : a color atlas / Kelvin Y.C. Teo,
    Wong Chee Wai, Andrew S.H. Tsai, Daniel S.W. Ting, Dan Milea ;
    edited by Shu Yen Lee and Gemmy C.M. Cheung ; foreword by Wong Tien Yin.
Description: Second edition. | Singapore : World Scientific, 2018. |
    Includes bibliographical references and index.
Identifiers: LCCN 2018017935| ISBN 9789813271777 (hardcover : alk. paper) |
    ISBN 9813271779 (hardcover : alk. paper) |
    ISBN 9789813239562 (pbk. : alk. paper) | ISBN 9813239565 (pbk. : alk. paper)
Subjects: | MESH: Eye Diseases--diagnosis |
    Tomography, Optical Coherence--methods | Atlases
Classification: LCC RE551 | NLM WW 17 | DDC 617.7/350757--dc23
LC record available at https://lccn.loc.gov/2018017935

British Library Cataloguing-in-Publication Data
A catalogue record for this book is available from the British Library.

For any available supplementary material, please visit
http://www.worldscientific.com/worldscibooks/10.1142/10972#t=suppl

LIST OF CONTRIBUTORS AND EDITORS

Authors

Kelvin YC TEO
MBBS, MMed (Ophth)
Associate Consultant
Singapore National Eye Centre

Chee Wai WONG
MBBS, MMed (Ophth), MCI, FAMS
Consultant
Singapore National Eye Centre

Andrew SH TSAI
MBBS, MMed (Ophth), FRCOphth, FAMS
Associate Consultant
Singapore National Eye Centre

Daniel SW TING
MBBS (1st Hons), BMedSci, MMed (Ophth), FAMS, PhD
Associate Consultant
Singapore National Eye Centre
Assistant Professor
Duke NUS Medical School

Dan MILEA
MD, PhD
Senior Clinician, Neuro-ophthalmology Department,
Singapore National Eye Centre
Head, Visual Neuroscience Group, Singapore Eye Research Institute
Professor, Duke NUS Medical School

Editors

Gemmy CM CHEUNG
FRCOphth, FAMS
Senior Consultant & Deputy Head, Medical Retina Service
Head, Ophthalmic Imaging Service, Singapore National Eye Centre
Senior Clinician Investigator & Head, Retina Research Group
Singapore Eye Research Institute
Associate Professor, Duke-NUS Graduate Medical School
National University of Singapore

Shu Yen LEE
MBBS, MMed (Ophth), FRCS (Ed), FAMS
Senior Consultant & Deputy Head, Surgical Retina
Clinical Director, Retina Centre, Singapore National Eye Centre
Adjunct Associate Professor, Duke-NUS Graduate Medical School
National University of Singapore

ACKNOWLEDGMENTS

We are extremely grateful to the Ophthalmic imaging team at Singapore National Eye Centre for supplying the images without which this book could not have been written:

Joseph HO
Principal Ophthalmic Imaging Specialist, Singapore National Eye Centre

Paul CHUA
Senior Ophthalmic Illustration Specialist, Singapore National Eye Centre

Kasi SANDHANAM
Ophthalmic Imaging Specialist, Singapore National Eye Centre

Jackson KWOK
Ophthalmic Imaging Specialist, Singapore National Eye Centre
Vitreo-Retinal Department, Singapore National Eye Centre
Topcon Singapore Medical

FOREWORD

Ocular imaging has revolutionized the diagnosis and management of eye diseases.

Since its inception a decade ago, optical coherence tomography (OCT) has had a transformative impact on all aspects of ophthalmology, providing quick, high resolution and non-invasive method of imaging for almost all parts of the eye. OCT has allowed a greater understanding of the normal anatomy and physiology of the eye and the pathophysiology of major ocular diseases. The clinical impact of OCT is astronomical, and it is now the most commonly used imaging modality in ophthalmology with over 30 million scans performed worldwide yearly. This demand has fueled the need for even faster and better quality scan resulting in the rapid evolution of OCT technology from the first-generation time domain OCT to high-resolution spectral domain OCT in a few short years.

A new generation of OCT, the swept source OCT (SS-OCT), allows us to acquire higher resolution images coupled with the ability to scan deeper into the eye at faster scan speeds. Concurrently, with the advent of OCT angiography (OCT-A), we are also able to image the microcirculation of the retina by composing the motion signals of moving blood cells. OCT-A and in particular SS-OCTA represents the newest iteration of OCT technology. OCT-A marks a progression from previous dye-based angiography which is invasive, carries a risk of anaphylaxis and is often time-consuming and uncomfortable for our patients. OCT-A also allows us to examine individual "slabs" of retina circulation where previous dye-based techniques could not.

SS-OCTA has the distinct advantage over other OCT platforms in that it can image the choroidal vasculature providing a greater understanding of how the choroidal circulation affects retina disease such as in cases of age-related macular degeneration and central serous chorioretinopathy. SS-OCTA is still a technology in its infancy but holds great potential in further improving our understanding of ocular disease to help guide our management options.

The SNEC Retina Department, in conjunction with the SNEC Ocular Imaging Department, showcased the structural SS-OCT scans of common retina conditions in the first version of this Atlas. We are pleased to present a new version of this Atlas which is a collection of images and annotations describing the microanatomical structures, as well as OCT-A features of the diseased eye captured on the SS-OCT.

This Altas, like its predecessor, serves as a practical guide for retina specialists and clinicians with a particular interest in retinal diseases. A new chapter also showcases the use of SS-OCT and SS-OCTA in the imaging of the optic nerve relevant to the sub-specialty of neuro-ophthalmology.

We hope that this book will inspire further research and application in this new exciting field of ocular imaging.

Professor Wong Tien Yin

Medical Director, Singapore National Eye Centre

Vice-Dean, Office of Academic & Clinical Development
Duke-NUS Medical School
National University of Singapore

Deputy Group CEO (Research and Education)
Singapore Health Services (SingHealth)

CONTENTS

1A Introduction to Swept-Source Optical Coherence Tomography

Optical coherence tomography (OCT) is an *in vivo* non-invasive patient friendly modality for visualizing ocular structures. It produces high resolution, cross-sectional images of posterior segment structures, akin to an optical biopsy.

In current ophthalmic practice, the OCT has emerged as an important ancillary test. It helps the ophthalmologist in the diagnosis of various vitreo-retinal conditions and in monitoring response to treatment on follow up. The expanding role of OCT has resulted in a reduction in the number of fundal fluorescein angiography ordered.

Based on the principle of optical reflectometry, it measures the intensity and echo time delay of light that is scattered from the tissues of interest. Light from a broadband light source is broken into a reference arm and a sample arm that is reflected from structures at various depths within the posterior pole of the eye. Backscattered light can be detected via time domain or Fourier domain methods of detection. In this book, we focus on the latest swept-source OCT (SS-OCT) technology which is a form of Fourier domain detection.

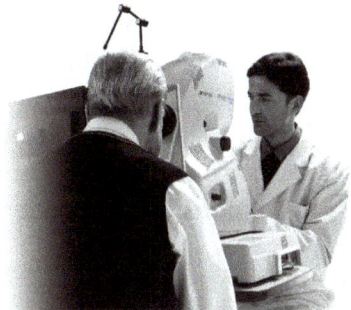

In SS-OCT scanning, the light source is rapidly swept in wavelength and the spectral interference pattern is detected on a single or a small number of receivers as a function of time. The spectral interference patterns obtained as a function of time then undergo a reverse Fourier transformation to generate an A-scan image.

The features and advantages of SS-OCT include:

- A large examination field (43°) which allows for simultaneous study of the macular area and optic nerve.
- It utilizes a longer wavelength (1050 nm) as compared to previous spectral domain technology (800 nm). This allows for deep range imaging which penetrates deeper to visualize the choroid and sclera in detail. It is also able to better image through media opacities, compared to its predecessors.
- An invisible scanning line contributes to reduced patient eye motion, enhancing successful rates of scanning and fast examination workflow.
- A fast scanning speed of 100,000 A scans per second, and faster imaging acquisition time. It takes approximately 0.01 second to obtain a B-scan, and 0.9 seconds for a 3-dimensional (3D) scan.

In clinical practice, the use of SS-OCT has also allowed for better visualization of vitreous anatomy and the vitreo-retinal macular interface. The inner and outer retinal layers are also more clearly defined. From a SS-OCT image, one can decipher the inner layers and the ganglion cell complex which comprises the inner plexiform layer, ganglion cell layer and nerve fiber layer. The outer retinal layers of the normal eye also shows three distinct bands: (a) retinal pigment epithelium (RPE) band which consists of the RPE, Bruch's membrane and choriocapilarries; (b) anterior to RPE band which comprises the outer limiting membrane, inner segment-outer segment line and Verhoeff's membrane; (c) posterior to RPE band which consists of the middle and outer layers of the choroid.

Other possible examination which can be performed include:

- Transverse or "enface" OCT. Some possible uses include visualizing the vitreo-macular interface before and after macular surgery, and to analyze different patterns of the choroidal neovascular network or polypoidal choriodal vasculopathy.

- Stereo photographs of fundus can be obtained with the green filter.
- 3D SS-OCT system allows visualization of the whole retina in a 12 × 9 mm cube.

Different Modes

1. Analysis Feature

Advanced layer detection algorithm detects 7 layers of the retina. Thickness map and caliper function allow for detailed analysis. Other features include automated choroid-scleral interface detection and choroid map analysis. Ganglion cell layer analysis and map are useful for detection and follow up for glaucoma. Generation of superpixels map and comparison with normative database also allows for glaucoma diagnosis.

2. Import Function

The SS-OCT system has the ability to import color photos, fluorescein angiography, indocyanine green angiography and fundus autofluorescence images and compare with OCT images. Image overlay is made possible by recognition and alignment of retinal vessels.

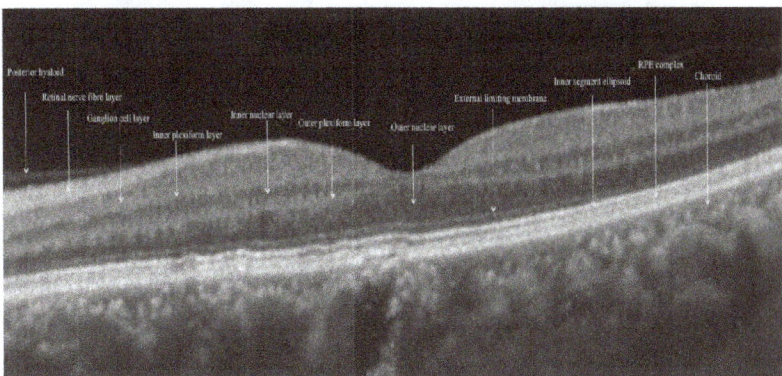

Fig. 1A.1.   Layers of the normal retina as seen on DRI OCT.

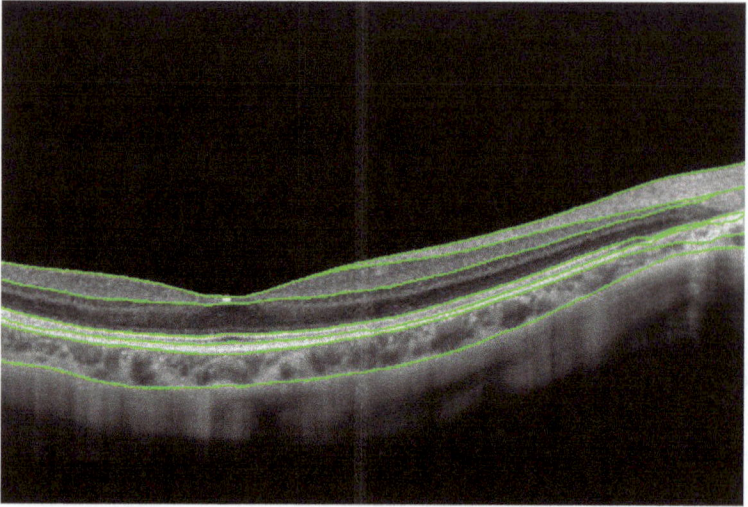

Fig. 1A.2.   7-layer detection.

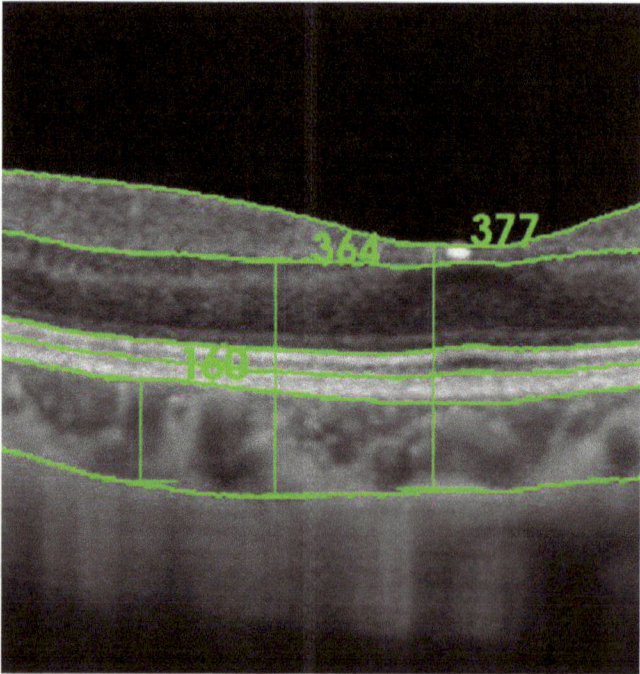

Fig. 1A.3.   Caliper function.

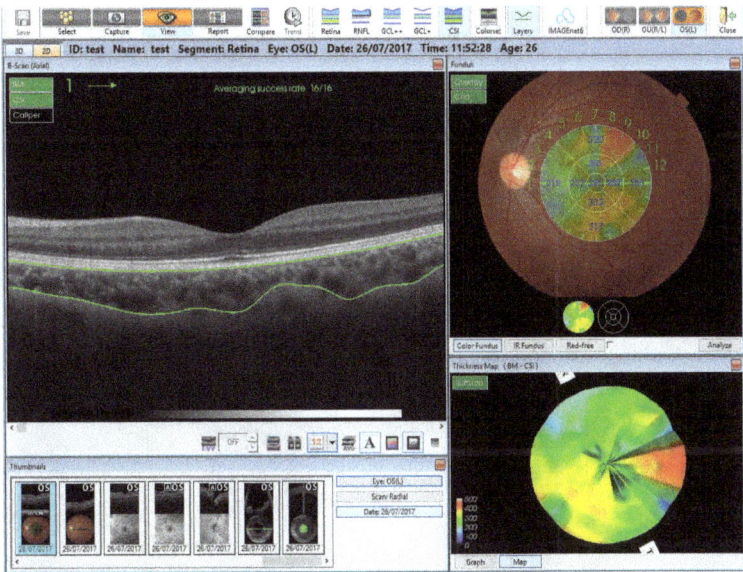

Fig. 1A.4. Automated choroid-scleral interface (CSI) detection. Top-right picture shows choroidal thickness (µm). Bottom-right picture shows the color-coded choroid thickness map.

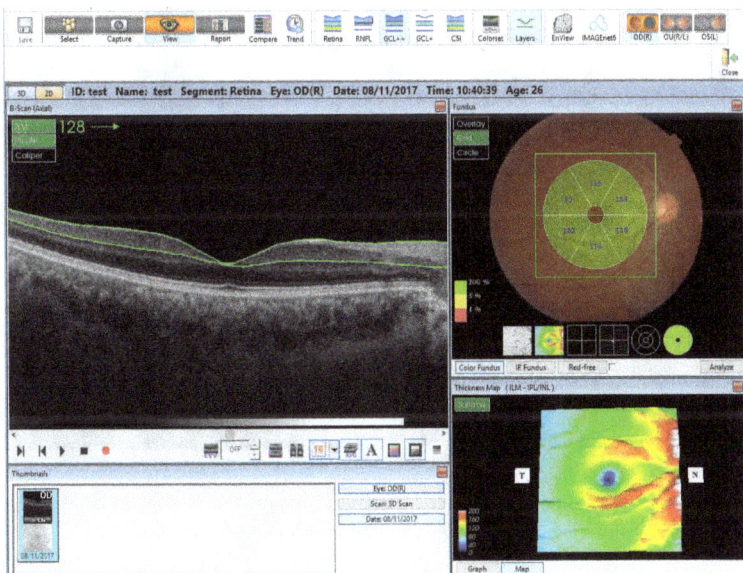

Fig. 1A.5. Automated ganglion cell layer/retinal nerve fiber layer segmentation allows for early detection of structural glaucoma defects.

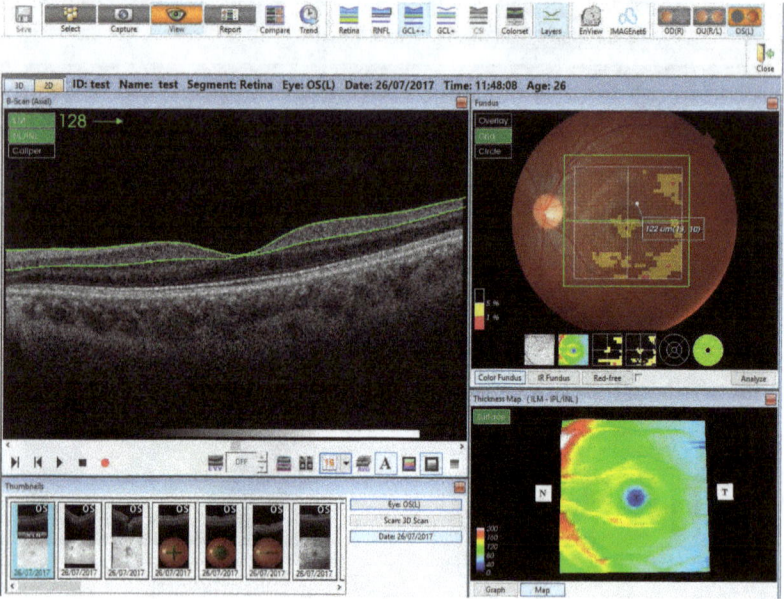

Fig. 1A.6. Superpixel feature (top right photo) showing the ganglion cell/retinal nerve fiber layer thickness in a particular grid.

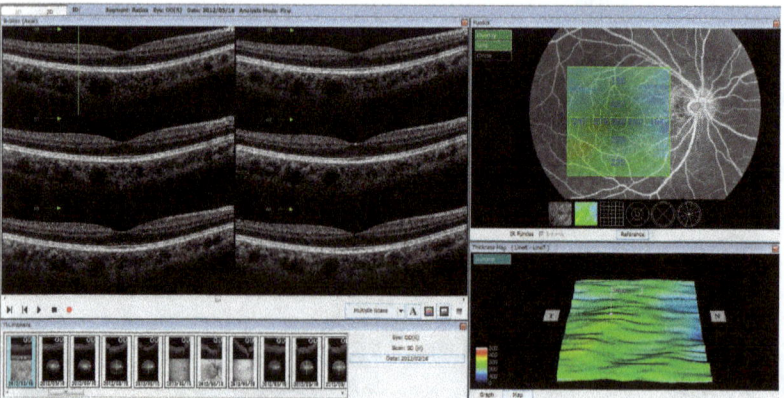

Fig. 1A.7. Import function — the numerical values on the overlay in the top-right photo refers to the thickness (μm) of the segmentation layer selected.

3. 3D Volume Rendering

Captured 3D image can be cropped or peeled.

4. Follow-up Exam and Comparison Function

Inbuilt software detects the same scanning location of the fundus image during follow-up examination, which allows for comparison (Fig. 1A.8) and monitoring of response to treatment.

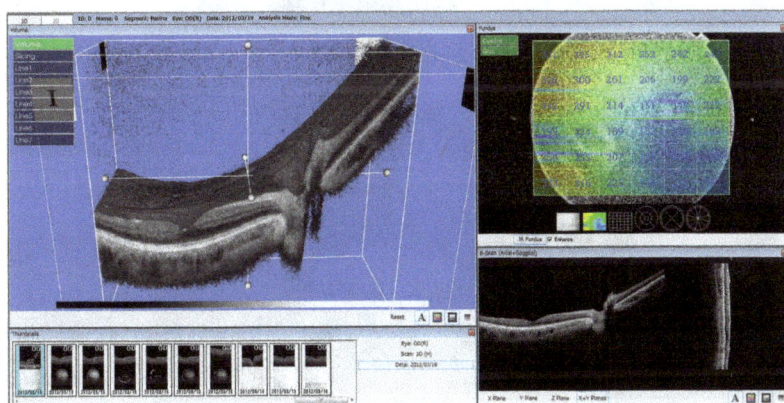

Fig. 1A.8. Cropping — the numerical values in each grid in the top-right photo refers to the thickness (μm) of the segmentation layer selected.

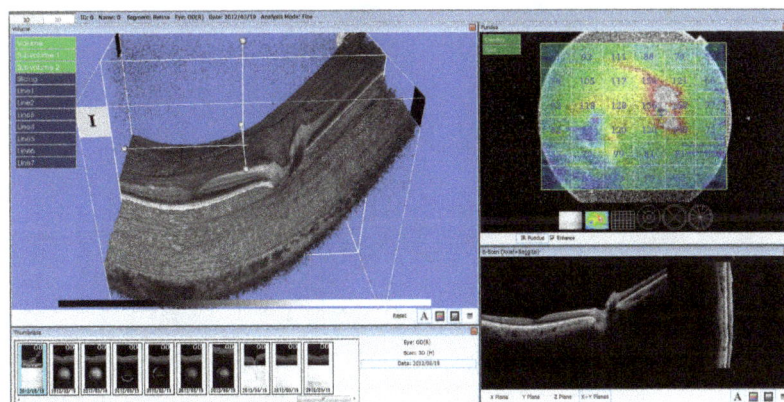

Fig. 1A.9. Peeling.

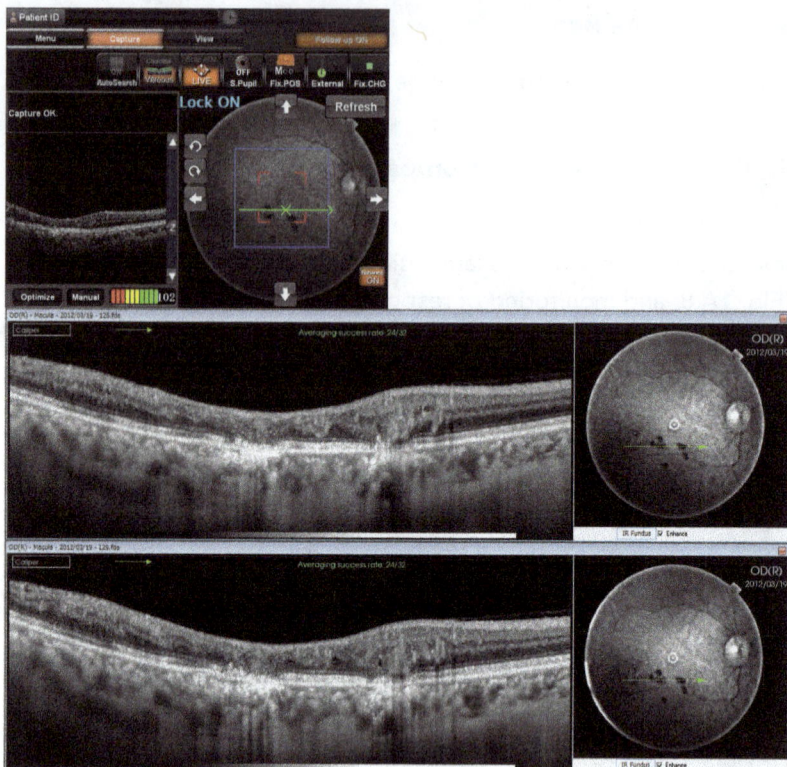

Fig.1A.10.   Comparison.

1B Introduction to Optical Coherence Tomography Angiography

Optical coherence tomography angiography (OCT-A) is a dye free, non-invasive method of visualizing the microvasculature of the retina in three dimensions.[1,2] The principle behind OCT-A technology is based on the detection of differences in either amplitude, phase variance or intensity between serial B scans performed at the same location on the retina. These multiple scans are registered and the degree of decorrelation in signal is calculated and displayed graphically. Stationary tissue produces a nearly constant reflection on OCT, whereas moving tissue produces signals that change over time. Pixels from stationary non-changing structures are displayed as black and those pixels which show changes or fluctuations are displayed as bright. When scans across several regions are combined, a three-dimensional data set is obtained[3] and various algorithms such calculation of speckled variance, split-spectrum amplitude decorrelation angiography (SSADA) and the optical micro-angiography (OMAG) algorithm[4,5] can be applied to reconstruct enface angiograms of the retina.

OCT-A is often compared with the traditional forms of angiography in the eye. Fluorescein (FA) and indocyanine green angiography (ICGA) are dye-based angiographic techniques that have been used to evaluate retina vasculature and various pathologies associated with vascular changes. The comparison is summarized in Table 1. Dye-based angiographic techniques have the advantage of detecting

Table 1.   Comparison between OCT-A and Dye-based Angiography.

	OCT-A	Dye Angiography
PROS	Non-invasive	Invasive
	No dye used	Dye based
	No side effects	Side effects of dye — anyphylaxis
	Image acquisition time: 2–5 min	Image acquisition time: 5–30 min
	High resolution	Low resolution
	Three dimensional — able to appreciate vasculature in different segments	Two dimensional — imaged in entirety, unable to discern deeper vasculature
CONS	Static flow information only	Dynamic flow information
	Artefact may hamper interpretation	Less artefact
	Small field of view	Large field of view with wide angle viewing options
	No stereoscopic function	Stereoscopic function

leakage which is a marker of vascular permeability and can be an important sign in certain disease process.[6] Leakage is a process of diffusion, and cannot be detected on OCT-A; however, images obtained on OCT-A are conversely not obscured by leakage, and hence involved abnormal vasculature can be easily visualized. OCT-A, unlike dye-based angiography is also a non-invasive process and can be performed repeatedly at every patient visit for disease monitoring. In addition, since OCT-A is derived from cross-section structural OCT, the vasculature at different depths of the retina can be imaged when the three-dimensional volumetric information is processed.[7] Commonly, the retina is segmented automatically by in-built software into the inner (or superficial) retina, deep retina, outer retina and choriocapillaris layers. The inner retina slab on OCT-A correlates well to the retina vasculature seen on FA; however, the deep plexus which corresponds to the deep retina layer cannot be seen on FA. The ability to resolve vasculature at this level on OCT-A is because specific layers can be selected and the overlying vessels do not obscure the underlying ones. This visualization of layers is useful in certain conditions; however, the interpretation of these retinal slabs should be performed cautiously to ensure accuracy of each slab segmentation so as not to mistake an artefact for pathology. Other shortcomings of OCT-A include flow sensitivity — very slow flow may not be detected and various artefacts generated by motion and projections contribute to the challenges of interpretation.[8] Nonetheless, various algorithms have been developed to overcome these shortcomings.

The Triton swept-source OCT-A (SS–OCTA) combines the use of swept-source OCT and a novel motion contrast measurement algorithm to provide high resolution images. The SS–OCT operates at 100Hz A line rate and a one-micron wavelength light source, allowing for deeper penetration to image the retina and choroid without loss of resolution. An additional advantage is that the SS–OCT technology can also provide a longer scan length as compared to spectral domain OCT (SD–OCT) imaging more of the retina at one pass. The proprietary motion contrast algorithm, named OCTARA (OCT angiography ratio analysis) uses a ratio-based result between each corresponding pixel in each corresponding scan while maintaining the full spectrum, hence preserving the axial resolution of each scan. Instead of amplitude decorrelation, this method uses the intensity ratio, calculated by the relative change between signal amplitude. This optimizes the angiographic visualization in both the retina and choroid and also enhances the minimum detectable signal relative to other decorrelation methods. Motion artefact is reduced by selectively averaging over multiple B scans.[9]

IMAGEnet6

This web-based software incorporates the visualization of angiographic datasets providing for both automated and customizable enface and cross-sectional views of OCT-A in conjunction with structural OCT.

The IMAGEnet6 web based software displays OCT-A information for each eye on a single screen to facilitate interpretation. The top four images bracketed in different colors display preset automated segmented retina/choroidal slabs. From left to right, the superficial retinal slab (orange), deep retinal slab (green), outer retinal slab (light blue) and the choriocapillaris slab (dark blue) are shown. The superficial retina slab show a vascular pattern that is similar to that seen in a traditional FA. The structural cross-sectional OCT-B scan is

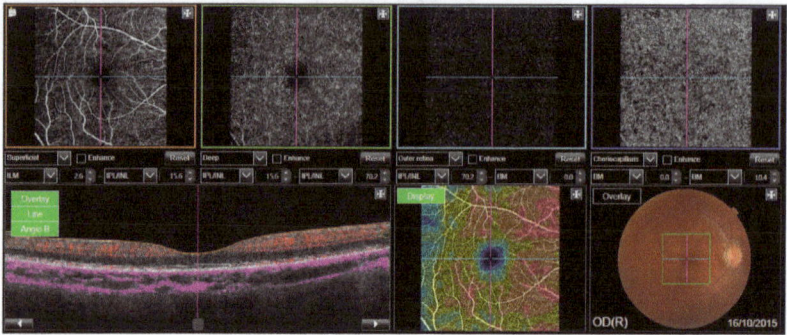

Fig. 1B.1.   OCT-A of a normal eye.

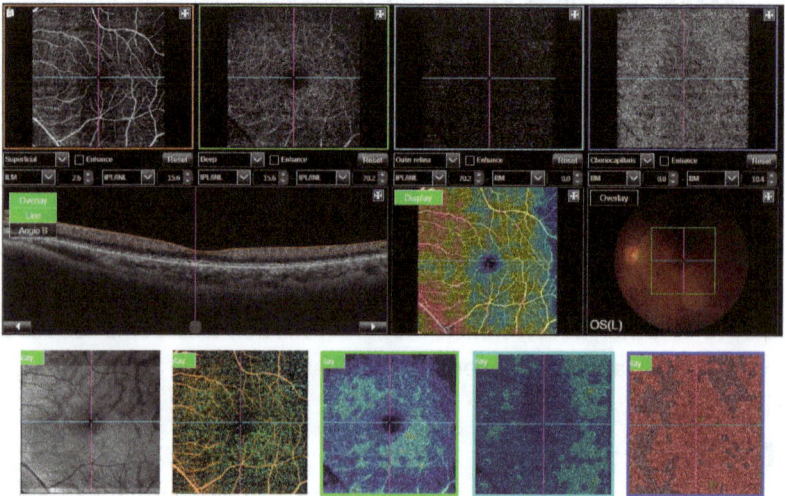

Fig. 1B.2.   OCT-A scan of normal eye (6 × 6 mm scan).

displayed in the bottom left panel with the option to toggle the angio B function (displayed), providing an overlay of areas of movement, correlating to blood flow within the scan. Red and purple colors on angio B scan representing areas of flow and interpretation of OCT-A enface images should always be correlated with this. In addition, the slab segmentation is also displayed as lines in the corresponding colors (orange line showing superficial slab selection).

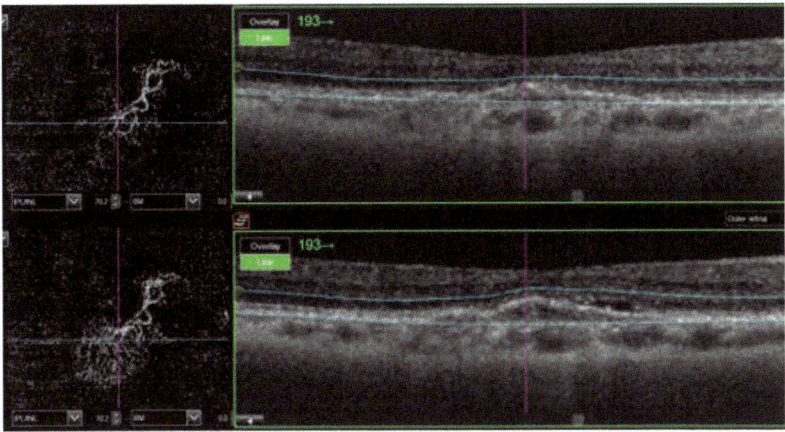

Fig. 1B.3. Inter visit registration.

The Triton is able to acquire OCT-A scans at 6 × 6 mm at 512 × 512 pixels. Similar display to Fig. 1B.1 is shown except that the structural B scan is displayed instead of the angio B scan in this image. The image to the right of the B scan can be toggled to show different displays: In the bottom row, from left to right, the enface structural scan, composite enface OCT-A, and density map of the deep, outer and choriocapillaris are shown (outline colors denoting each layer of the density map scans). This current display is the vessel density map of the superficial retinal slab, where denser vasculature is represented by warmer colors (red, orange) and less dense areas are represented by cooler colors (green, blue). The software also has the ability to manually select the segmentation lines. This is useful in the diseased eye where automated segmentation may often not provide accurate slab selection.

The software is also able to performs inter-visit registration for comparison of OCT-A scans on a single screen. The bottom image pair shows an active CNV lesion on OCT-A (left) with corresponding structural B scan showing some subretinal fluid (right). The image pair above shows the reduction in the size of the CNV lesion with resolution of subretinal fluid after treatment. The inter-visit registration allows for precise interpretation of the progress the CNV lesion showing the corresponding OCT-A as well as structural OCT-B scan from visit to visit.

References

1. Gao SS, *et al.* (2016). Optical coherence tomography angiography. *Invest Ophthalmol Vis Sci* **57**(9): 27–36.
2. Zhang A, *et al.* (2015). Methods and algorithms for optical coherence tomography-based angiography: A review and comparison. *J Biomed Opt* **20**(10): 100901.
3. Kashani AH, *et al.* (2017). Optical coherence tomography angiography: A comprehensive review of current methods and clinical applications. *Prog Retin Eye Res*, **60**: 66–100.
4. Jia Y, *et al.* (2012). Split-spectrum amplitude-decorrelation angiography with optical coherence tomography. *Opt Express* **20**(4): 4710–25.
5. Huang Y, *et al.* (2014). Swept-source OCT angiography of the retinal vasculature using intensity differentiation-based optical microangiography algorithms. *Ophthalmic Surg Lasers Imaging Retina* **45**(5): 382–9.
6. Johnson RN, *et al.* (2012). Fluorescein angiography: Basic principles and interpretation, in *Retina Fifth Edition*. Elsevier Inc.
7. Spaide RF, JM Klancnik Jr. and MJ Cooney. (2015). Retinal vascular layers imaged by fluorescein angiography and optical coherence tomography angiography. *JAMA Ophthalmol* **133**(1): 45–50.
8. de Carlo TE, *et al.* (2015). *A review of optical coherence tomography angiography (OCTA). Int J Retina Vitreous* **1**: 5.
9. Stanga P, *et al.* (2016). Swept-Source Optical Coherence Tomography Angio™ (Topcon Corp, Japan): Technology Review, in *OCT Angiography in Retinal and Macular Diseases. Karger Publishers* pp. 13–17.

2 Retinal Vascular Disease

Macroaneurysm

Retinal macroaneurysms are focal fusiform or saccular dilations, typically occurring within the first three bifurcations of the central retinal artery. They are seen predominantly in the elderly above 60 years of age and females are three times more likely to be affected. They are most commonly associated with systemic hypertension and generalized arteriosclerotic disease. Visual loss can be a result of macular exudation and/or hemorrhage into the vitreous, subretinal and intraretinal space.

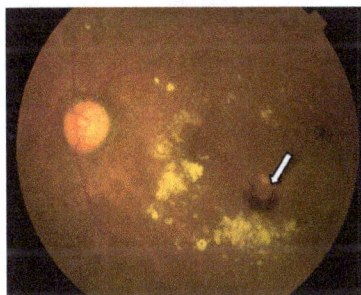

Fig. 2.1. Color fundus photograph of a patient with macroaneurysm (arrow), with surrounding intraretinal hemorrhage, hard exudates and macular edema.

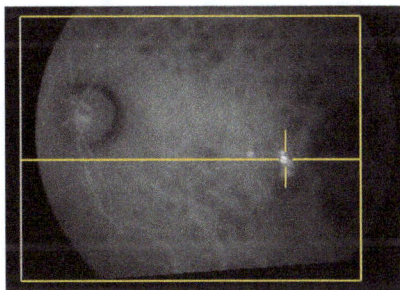

Fig. 2.2. ICG shows a hot spot corresponding to the position of the macroaneurysm on fundus photograph.

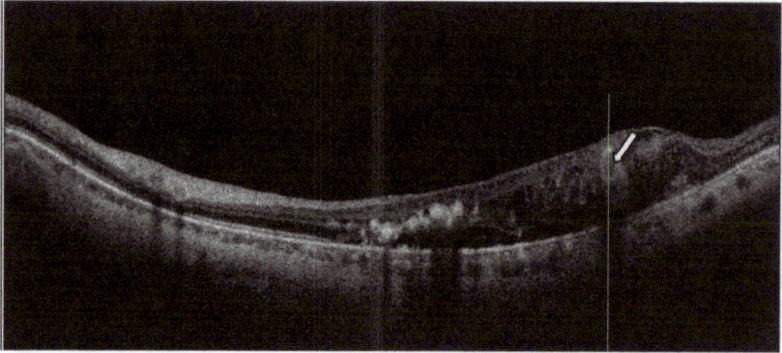

Fig. 2.3.   SS-OCT of the above patient. A hyperreflective lesion corresponding to the macroaneurysm is seen within the intraretinal space (arrow).

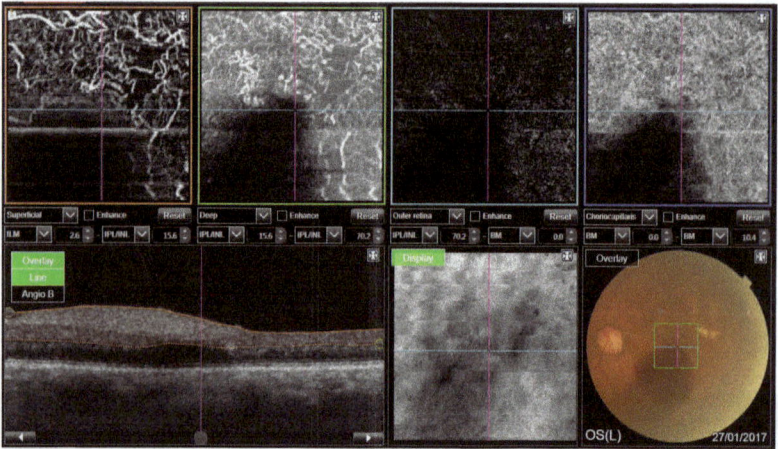

Fig. 2.4.   3 × 3 mm OCT-A of a patient with left eye retinal macroaneurysm. The vasculature in all the layers are obscured at the region of hemorrhage. The superficial and deep vascular shows disorganized and tortuous vessels.

Idiopathic Macular Telangiectasia

Idiopathic macular telangiectasia is a group of retinal vascular disorders characterized by retinal capillary dilatation of unknown cause affecting the macula. Macular telangiectasia Type 1 is an aneurysmal telangiectasia associated with exudation, almost always unilateral and affecting men more commonly. Macular telangiectasia Type 2 is a bilateral disease involving the macular capillary network and characterized by atrophy of the neurosensory retina. It typically affects individuals in the fifth or sixth decade of life with no gender predilection. It can be further classified into a nonproliferative stage where there is foveal atrophy, and a proliferative stage with the development of subretinal neovascularization. Table 2.1 shows the stages of macular telangiectasia Type 2.

Table 2.1.   Idiopathic Juxtafoveal Retinal Telangiectasis of Gass and Blodi (based on Biomicroscopy and Stereoscopic Fluorescein Angiography).

Stage 1	No biomicroscopic abnormality, no or minimal capillary dilation, mild staining of outer perifoveal retina
Stage 2	Slight graying of perifoveolar retina, no or minimal biomicroscopically visible telangiectatic vessels, but capillary telangiectasis of outer capillary network temporally on fundus autofluorescence
Stage 3	One or several slightly dilated and blunted retinal venules descending into outer perifovea, typically temporally
Stage 4	Pigment hyperplasia, often surrounding right-angle venules
Stage 5	Subretinal neovascularization, often in proximity to intraretinal pigment migration

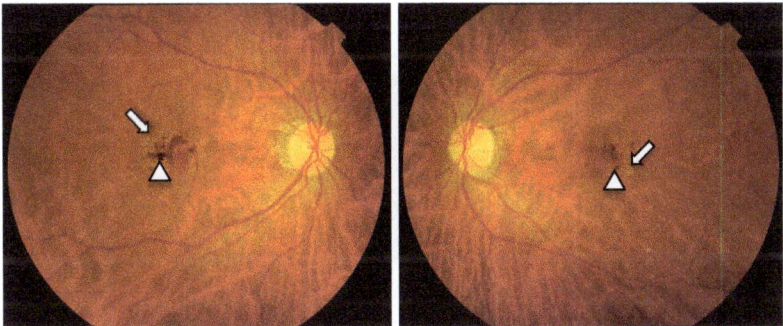

Fig. 2.5. Macular telangiectasia Type 2. Note the presence of bilateral disease affecting the macular, characterized by pigment proliferation (arrowheads) and crystalline deposits (arrows).

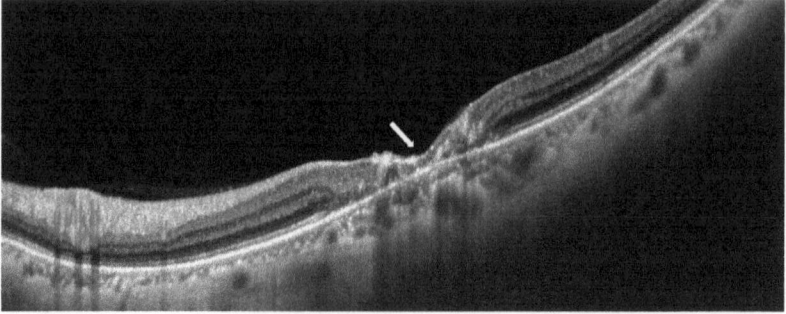

Fig. 2.6.   SS-OCT of the right eye shows foveal atrophy (arrow) with loss of the outer retina, representing a late stage of disease.

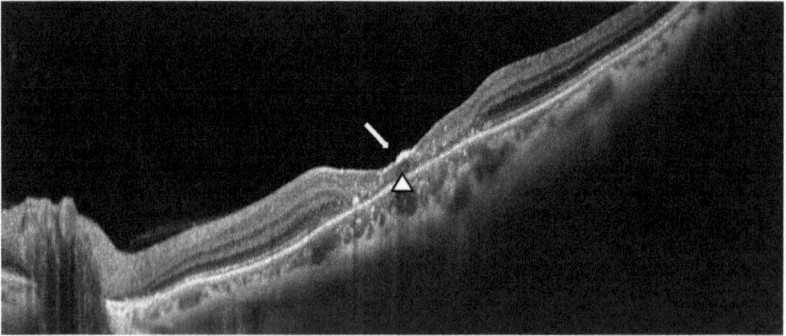

Fig. 2.7.   SS-OCT of the left eye shows foveal atrophy (arrow) and the presence of an intraretinal cyst (arrowhead).

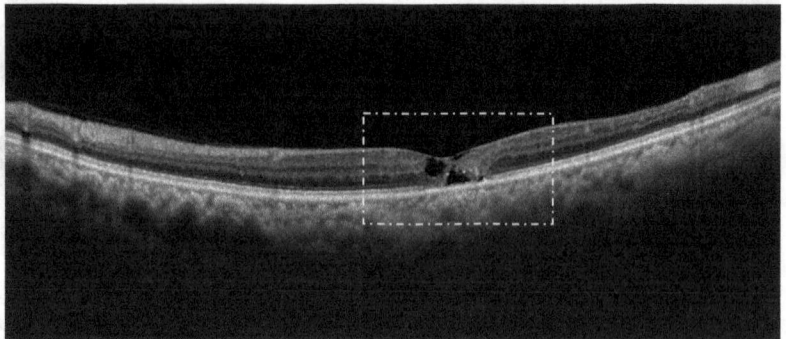

Fig. 2.8.   SS-OCT of another patient with macular telangiectasia demonstrating typical cystoid cavities in the absence of macular edema.

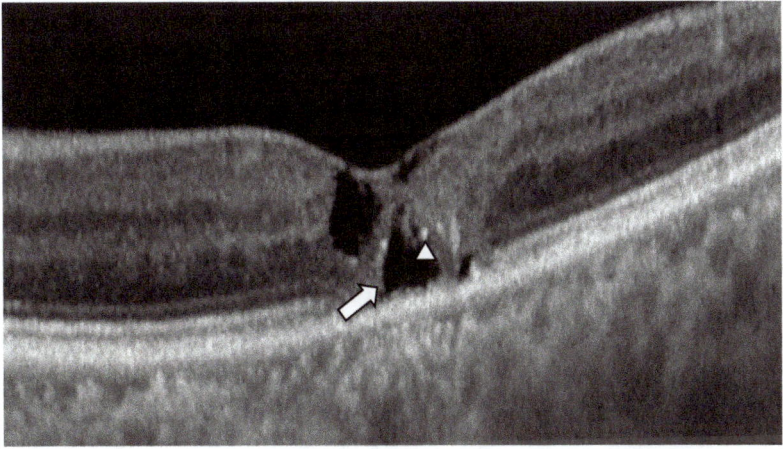

Fig. 2.9. Magnified view of the white box above. Note the disruption of the outer retinal layers (arrow). The cystoid cavities are thought to result from dysfunction or lost of muller cells. Crystalline deposits are seen as hyperreflective spots on SS-OCT (arrowhead).

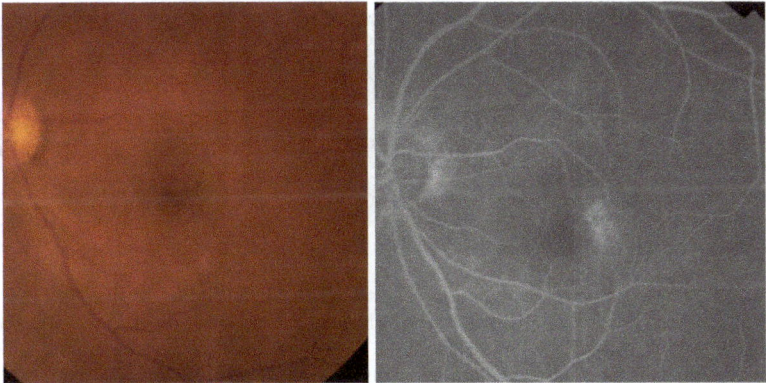

Fig. 2.10a. A 67-year old lady with stage 4 parafoveal telangiectasia of the left eye. Fundal fluorescein angiogram shows diffuse leakage from telangiectatic vessels in the late phase.

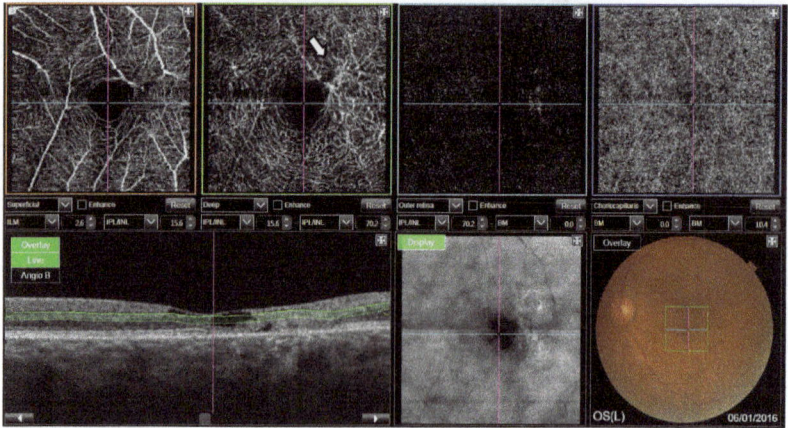

Fig. 2.10b. 3 × 3 mm OCT-A of the patient in Fig. 2.10a shows disruption and apparent widening of the foveal avascular zone (FAZ) due to intraretinal cysts. Telangiectatic vessels can be seen temporal to the FAZ in the deep capillary plexus (arrow). No choroidal neovascularization is seen in the avascular layer.

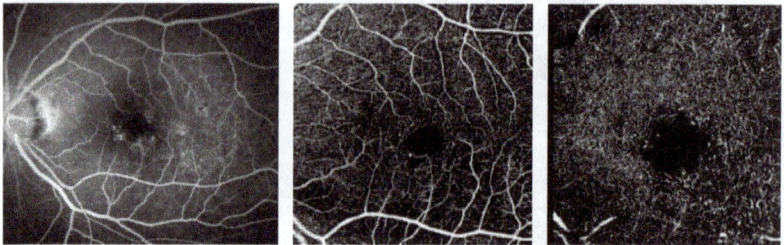

Fig. 2.11. A 67-year-old male with parafoveal telangiectasia of the left eye. Fluorescein angiogram (photo on left) shows leakage from telangiectatic vessels inferior and temporal to the foveal avascular zone. OCT-A of the superficial vascular plexus (middle photo) and deep vascular plexus (photo on right) shows some corresponding telangiectatic vessels. Vessels with slow flow may not be seen on OCT-A.

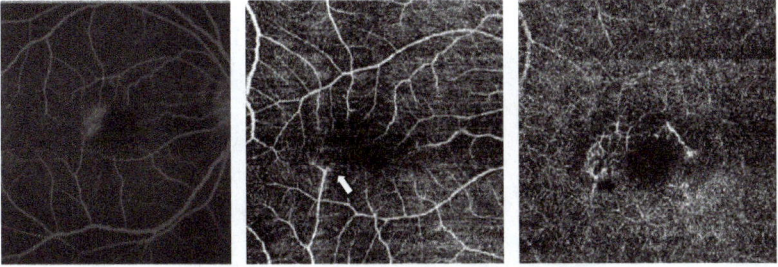

Fig. 2.12. A 67-year-old female with stage 3 parafoveal telangiectasia of the right eye. Fluorescein angiogram shows leakage from vessels temporal to the FAZ. There is blunting of the retinal vessels (arrow) inferotemporal to the fovea as seen on OCT-A of the superficial capillary plexus (middle photo). Telangiectatic vessels can be seen on OCT-A of the deep capillary plexus.

Retinal Vein Occlusion

Retinal vein occlusion occurs when flow obstruction occurs in either the central retinal vein or one of the branch retinal veins. It is most commonly seen as a complication of hypertensive retinopathy, where the arteriosclerotic retinal arterioles compress the retinal venules at an arteriovenous crossing. It is characterized by intraretinal hemorrhages and exudation in the areas of retina drained by the obstructed vessel. Decreased vision occurs when there is macular edema or significant ischemia. SS-OCT is valuable in visualizing the macula edema, and loss of ellipsoid zone which would suggest a poorer prognosis. Severe ischemia, if left untreated, can result in retinal neovascularization and/or neovascular glaucoma. Evidence from the BVOS and CVOS studies suggested that performing retinal laser photocoagulation to the area of capillary non perfusion (as seen on fluorescein angiography) can prevent neovascular complications such as vitreous hemorrhage and rubeosis. As shown in Figs. 2.12 and 2.13, OCT-A can also image these areas of capillary non perfusion. New evidence from Spaide RF[1] also suggested that damage to the deep vascular plexus as seen on OCT-A was associated with the recurrence of cystoid macular edema.

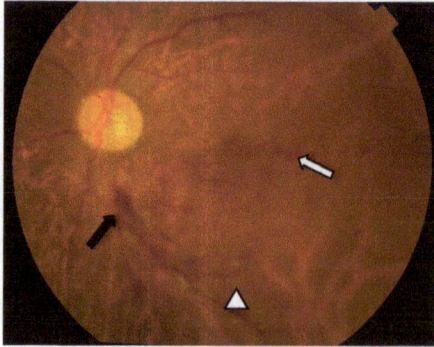

Fig. 2.13. Color photograph of an eye with inferotemporal branch retinal vein occlusion. Note the presence of intraretinal hemorrhage (white arrow) and cotton wool spots (white arrowhead) in the distribution of the occluded vessel and arteriovenous nipping (black arrow).

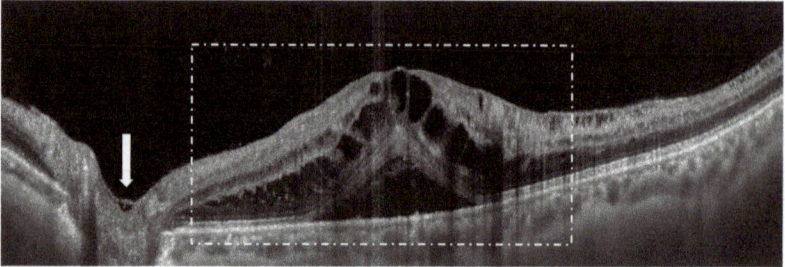

Fig. 2.14. SS-OCT of the above patient shows diffuse cystic intraretinal edema extending nasally to the optic disc (arrow).

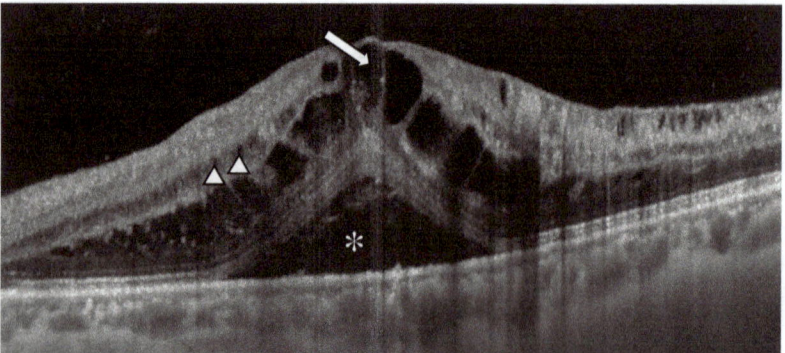

Fig. 2.15. Magnified view of the white box above. Multiple cystic spaces separated by bridging septae from fluid accumulation predominantly in the outer plexiform layer (arrow) can be seen clearly. Fluid can be seen in the subretinal space (asterisk) and inner plexiform layer (arrowhead).

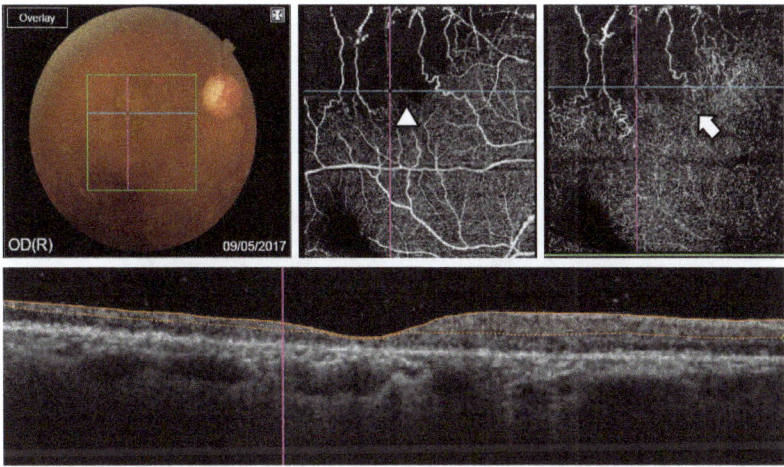

Fig. 2.16. A 65-year-old Chinese male with right eye ischemic superotemporal branch retinal vein occlusion. He initially presented with cystoid macular edema that was treated with intravitreal anti-vascular endothelial growth factor (anti-VEGF) agents. 3 × 3 mm OCT-A showed large areas of capillary non-perfusion in both the superficial (arrowhead) and deep capillary plexus (arrow). The corresponding B scan showed a dry macula with retinal thinning corresponding to the area of capillary non-perfusion.

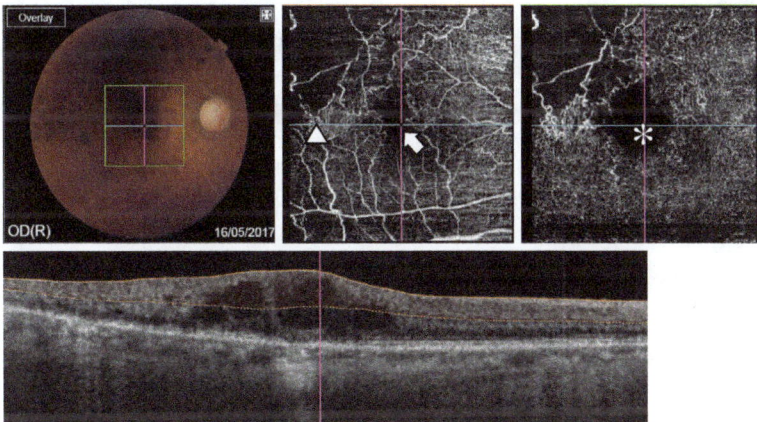

Fig. 2.17. A 76-year-old Chinese female with ischemic branch retinal vein occlusion of the right eye with cystoid macular edema. 3 × 3 mm OCT-A showed capillary non-perfusion at the superficial capillary plexus (arrowhead) and disruption foveal avascular zone (FAZ). In the deep capillary plexus, the foveal avascular zone also appears to be enlarged. However, the OCT-B scan showed presence of cystoid macular edema, which makes interpretation of the FAZ difficult.

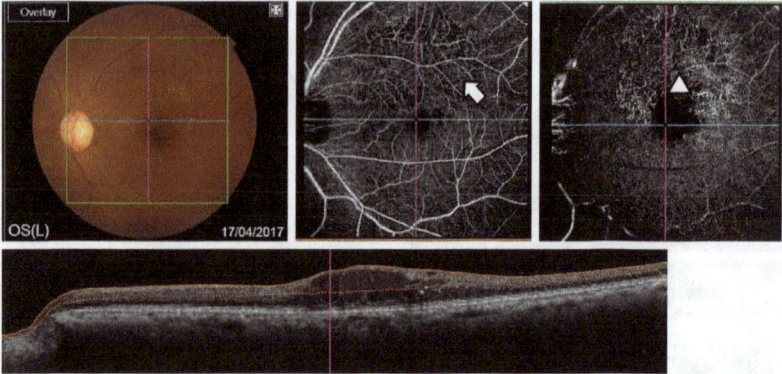

Fig. 2.18a.   A 61-year-old Chinese female with non-ischemic branch retinal vein occlusion of the left eye with cystoid macular edema. 6 × 6 mm OCT-A shows that the FAZ is relatively well preserved. Telangiectatic vessels can be seen in both the superficial capillary plexus (arrow) and deep capillary plexus (arrowhead). The corresponding OCT-B scan shows cystoid macular edema.

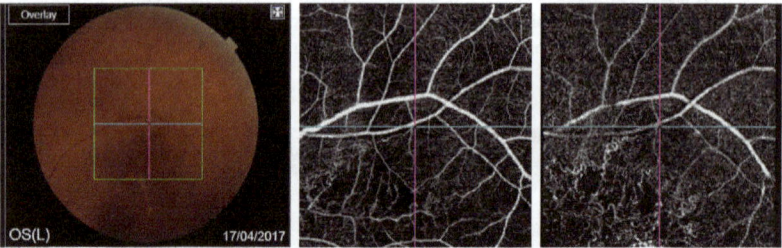

Fig. 2.18b.   The same patient in Fig. 18a. OCT-A centering on the vascular arcades shows that there is good capillary perfusion in both the superficial and deep plexus.

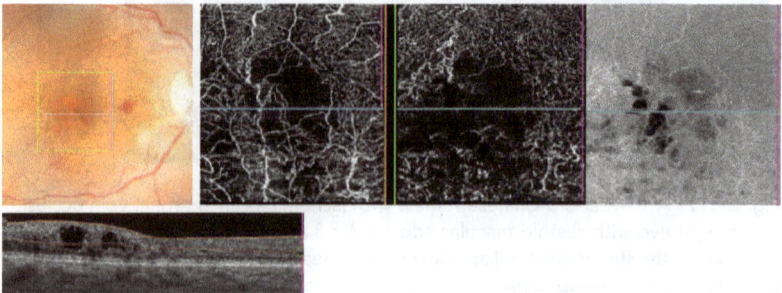

Fig. 2.19.   A patient with right macular branch retinal vein occlusion with cystoid macular edema. The enface OCT image (right–most) is useful for visualizing intraretinal cysts which can be confused for capillary non-perfusion on the OCT-A image (middle 2 images).

Diabetic Retinopathy

Diabetic retinopathy (DR), a diabetes-related, microvascular complication, is one of the leading causes for acquired visual loss. The retinal changes range from normal featureless fundus to a severe spectrum, consisting of neovascularization, vitreous hemorrhages, rubeosis iridis and glaucoma (Fig. 2.20).

It is, therefore, important to regularly screen for patients with diabetes early. We routinely screen patients using color fundus photos for DR, and OCT+/–angiogram (OCT-A) for diabetic macular edema (DME) that can be divided into center- and non-centered

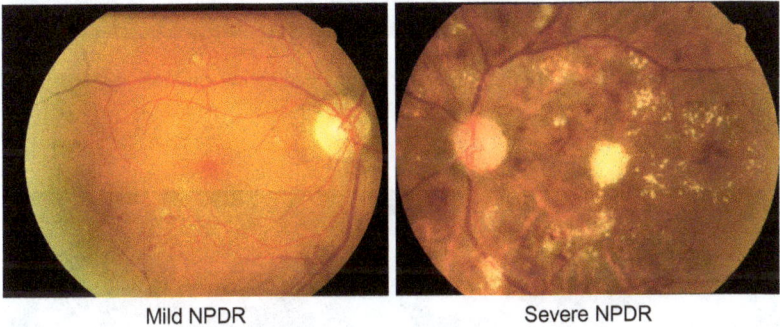

Mild NPDR   Severe NPDR

Fig. 2.20.   Fundus photos show right non-proliferative diabetic retinopathy.

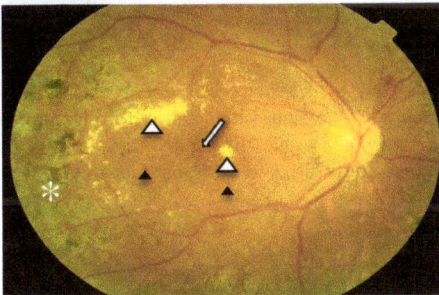

Fig. 2.21.   Color photograph of an eye with DME showing cystoid macular edema (arrow) and hard exudates (white arrowheads) from leaky microaneurysms (black arrowheads). Scars from previous laser photocoagulation can be seen (asterisk).

involvement. DME is a leading cause of visual impairment in patients with DR. It is characterized by leakage of fluid and lipids from retinal capillaries, resulting in cystic retinal swelling and lipid exudation. Previously treated with focal laser photocoagulation, anti-vascular

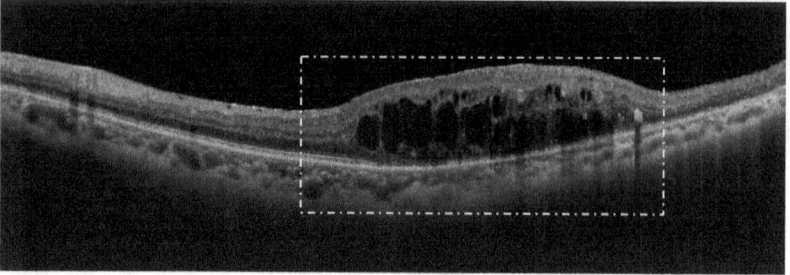

Fig. 2.22. SS-OCT of the above patient's right eye shows multiple cystic spaces, corresponding to that seen on the color photograph.

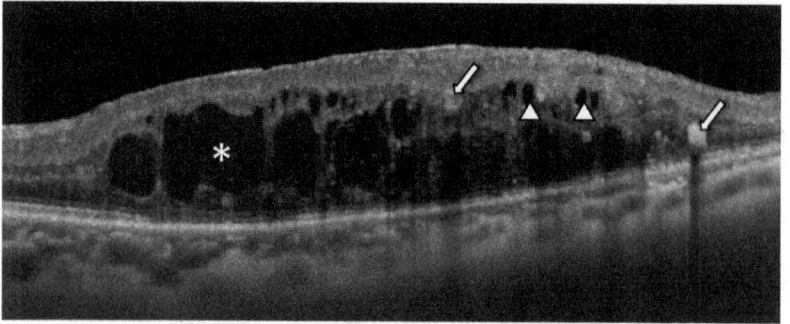

Fig. 2.23. Magnified view of the white box above. Large hyporeflective spaces (asterisk) with bridging septa, mainly within the outer plexiform layer can be seen. Small cystoid spaces are seen within the inner retina (arrowhead). Hard exudates appear as hyperreflective spots (arrows) with posterior shadowing.

endothelial growth factor treatment has emerged as the superior treatment option for visual improvement in recent years.

Until recently, OCT-A has enabled physicians to visualize the retinal microvasculature without the need for fundus fluorescein angiogram (FFA). Traditionally, FFA is ordered to ascertain the presence for retinal ischemia, macular edema/ischemia and neovascularization. With the advent of OCT, FFA is no longer ordered as a routine test, mainly for those patients with unexplained visual impairment. OCT-A enables the physicians to closely examine the superficial vascular plexus (SVP),[2] deep vascular plexus (DVP), and outer nuclear and choriocapillary layers. The OCT-A qualitative changes include: 1) microaneurysms (characterized by focally dilated saccular or fusiform capillaries); 2) macular ischemia (irregular FAZ, enlarged perivascular spaces, disruption of the

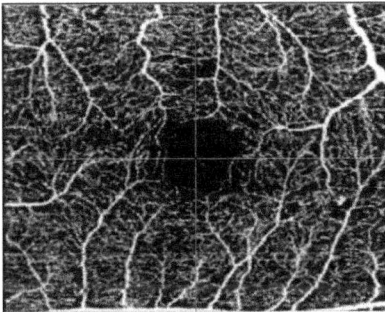

Moderate NPDR (SVP)   Moderate NPDR (DVP)

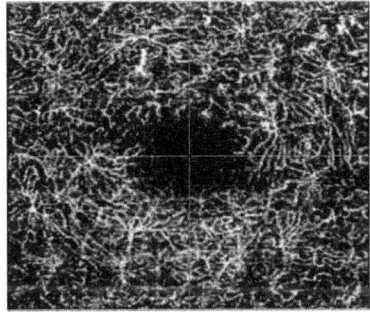

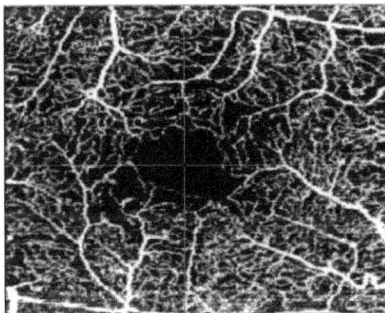

Proliferative DR (SVP)   Proliferative DR (DVP)

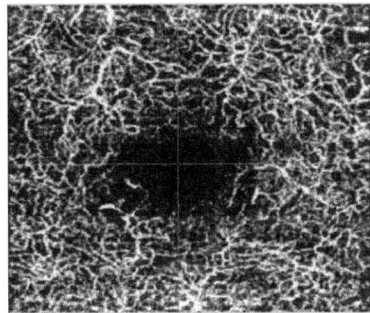

Fig. 2.24. The superficial vascular plexus (SVP) and deep vascular plexus (DVP) for moderate non-proliferative DR (NPDR) and proliferative DR.

capillary ring and non-perfusion); and 3) disc neovascularization. On the other hand, some quantitative tools were developed to assess the structural changes in diabetic eyes, including capillary density, fractal dimensions, vascular flow, FAZ areas, FAZ perimeters and diameters (horizontal and vertical), acircularity index (ratio of perimeter of FAZ to the perimeter of a circle with an equal area) and axis ratio (ratio between the major and minor axis of an ellipse).

As compared with the normal eyes without DME, the eyes with diabetic macular edema was shown to have more microaneurysms, lower vascular flow density and a larger FAZ area at the DVP layer. Moreover, these features were also co-related to poorer response to anti-VEGF injections. Given that cystic changes in DME may affect the segmentation and assessment of OCT-A at different layers, it is important to ensure the segmentation line is appropriately drawn (may need to be re-adjusted manually if necessary). Figure 2.24 shows the SVP and DVP of moderate NPDR and proliferative DR.

References

1. Spaide RF (2016). Volume-rendered optical coherence tomography of retinal vein occlusion pilot study. *Am J Ophthalmol* **165**: 133–44.
2. Ting DSW, Tan GSW, Agrawal R, Yanagi Y, Cheung CM and Wong TY (2017). Optical coherence tomography angiogram for diabetes and diabetic retinopathy. *JAMA Ophthalmology*. doi:10.1001/jamaophthalmol.2016.5877

3

Central Serous Chorioretinopathy

Central serous chorioretinopathy (CSCR) is a disease characterized by choroidal hyperpermeability and leakage of fluid through the retinal pigment epithelium. Fluid accumulation results in neurosensory retinal detachment and/or retinal pigment epithelium detachment.

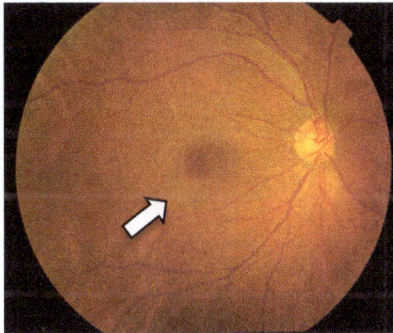

Fig. 3.1. Color photograph of an eye with CSCR showing neurosensory detachment of the central macula (white arrow).

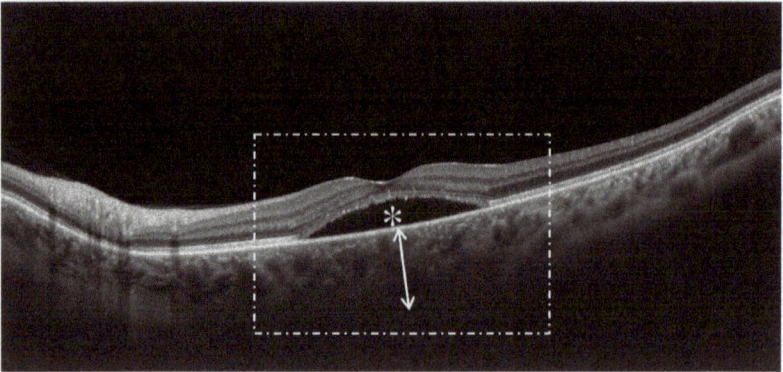

Fig. 3.2. SS-OCT shows a subfoveal dome-shaped hyporeflective space between the neurosensory retina and RPE, representing subretinal fluid (asterisk). Note the increased choroidal thickness (double headed arrow).

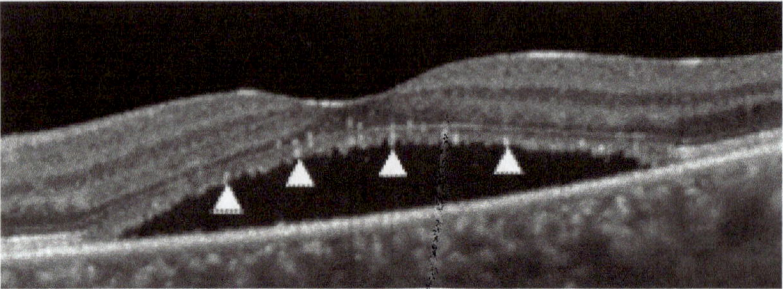

Fig. 3.3. Magnified view of the dashed box above: small hyperreflective spots represent lipofuscin deposits (white arrowheads), corresponding to the white spots on the color photograph.

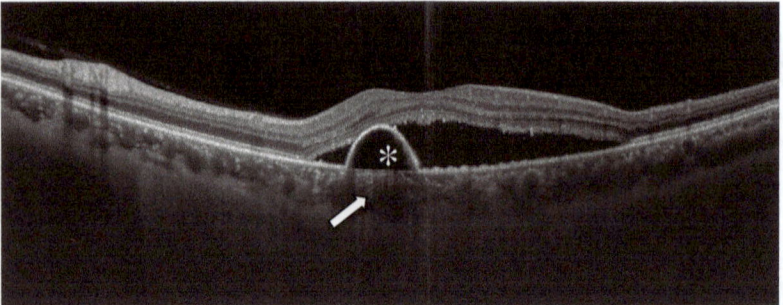

Fig. 3.4. Pigment epithelial detachment (asterisk) is commonly seen in CSCR. Note the spongy choroid (white arrow).

CNV in Central Serous Chorioretinopathy

Choroidal neovascularization (CNV) is an uncommon finding in central serous chorioretinopathy (CSCR), with incidence estimates ranging from 2% to 9%.[1] The presence of a CNV is a cause of reduced visual acuity in long-standing CSCR. There are several clinical features that are indicative of a CNV, including older patients with CSCR, eyes with diffused RPE loss, subretinal and/or sub-RPE hemorrhage, presence of lipid exudation, sub-RPE, subretinal or intraretinal fluid and subretinal hyperreflective material on OCT. Some of these features can also been seen in chronic CSCR which can make the diagnosis of CNV challenging. Even with the use of FA or ICGA, overlap features can also make the diagnosis of CNV difficult.[2] OCT-A provides a unique imaging modality in these cases, allowing for the visualization of a CNV network which subsequently alters clinical management. While CSCR alone may be treated conservatively, the presence of CNV may warrant anti-VEGF therapy.

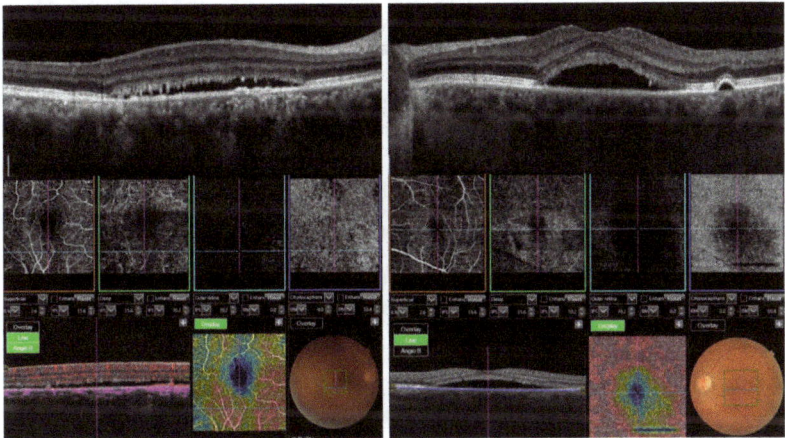

Fig. 3.5.   OCT-A in CSCR.
This series of images show 2 eyes with CSCR. OCT-A scans show no presence of any CNV lesion. CSCR in both these eyes resolved spontaneously.

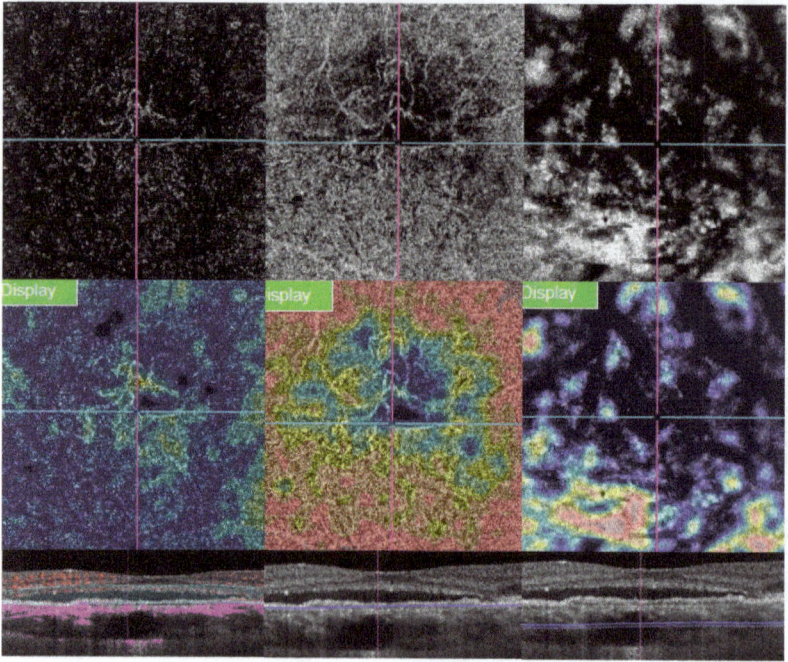

Fig. 3.6. OCT-A in detection of CNV in CSCR.
This series of images show the CNV in a case of chronic CSCR as a fine network of vessels in the outer retina slab (top left image). The projection of the CNV onto the choriocapillaris layer improves the visualization of the network (top middle image). Corresponding vessel density scans and B scans are shown. The right column shows the user-defined haller's layer slab which shows a large pachyvessel in the area that corresponds to the origin.

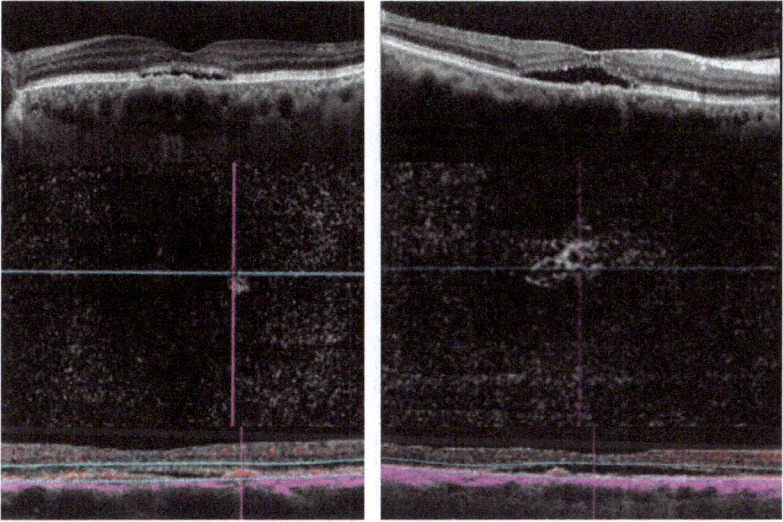

Fig. 3.7.   CNV in CSCR.
This series of images show the CNV in 2 eyes (left and right images) with what appears to be typical CSCR. The CNV lesions were detected on OCT-A (middle images), with corresponding angio B scans showing areas of motion (bottom row). Both eyes responded well after treatment with anti-VEGF therapy, resulting in the resolution of the CNV. Note the presence of the pachychoroidal vessels on the structural OCT-B scans (top images).

References

1. Gomolin JE. (1989). Choroidal neovascularization and central serous chorioretinopathy. *Can J Ophthalmol* **24**(1): 20–3.

2. Loo RH, et al. (2002). Factors associated with reduced visual acuity during long-term follow-up of patients with idiopathic central serous chorioretinopathy. *Retina* **22**(1): 19–24.

ЧА Age-related Macular Degeneration

Age-related macular degeneration (AMD) is one of the main causes of visual impairment in individuals over the age of 50 years. Neovascular AMD (nAMD) or exudative AMD is an advanced form of AMD and is the leading cause of acute visual loss related to AMD. Choroidal neovascularization (CNV) is the hallmark of nAMD and can include features like hemorrhage, fluid exudation, fibrosis and photoreceptor damage which ultimately leads to vision loss.[1] Treatment for this condition is with the use of anti-vascular endothelial growth factor (anti-VEGF) agents.[2]

AMD CNV is traditionally diagnosed on dye-based angiography such as fluorescein or indocyanine green (ICGA). Optical coherence tomography (OCT) is also an important structural imaging modality which can help with the diagnosis and monitoring of response to treatment.[3] The evaluation of AMD CNV is one of the most important applications of OCT-A. While it has not superseded dye-based angiographic techniques, it holds great promise as a non-invasive tool to diagnose AMD CNV and to assess its morphology and response to anti-VEGF therapy.

Age-related Macular Degeneration (Early)

The early form of age-related macular degeneration (AMD) is characterized by the presence of drusen in the macula. Drusens

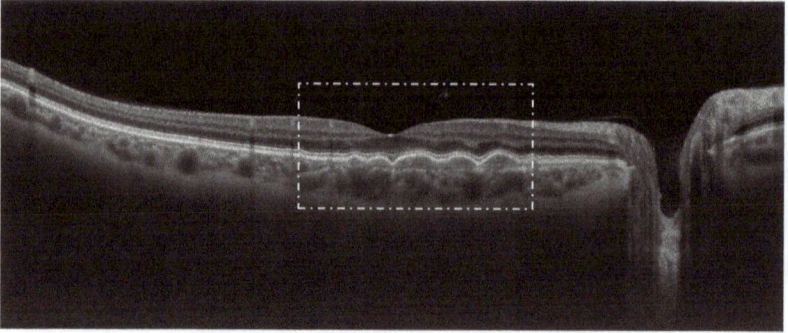

Fig. 4A.1.   SS-OCT of a patient with drusen.

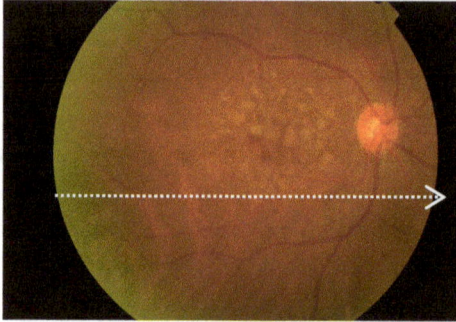

Fig. 4A.2.   Color fundus photograph of macula with drusenoid PED.

vary in size and characteristics. Soft indistinct drusen and pigment abnormalities are considered high risk characteristics for development of geographic atrophy or neovascular AMD. Color fundus photograph (Fig. 4A.1) shows macula of a patient with high risk drusens with corresponding SS-OCT foveal cut (Fig. 4A.1). The white dashed line indicates radial SS-OCT scan centered on the fovea with arrowheads indicating drusen with overlying undulating intact RPE. Debris can accumulate below RPE causing small detachments. These debris are referred to as drusenoid RPE detactments or drusenoid pigment

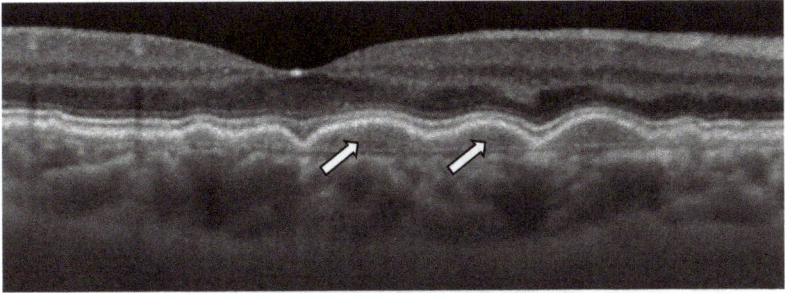

Fig. 4A.3.   Magnified view of white box in Fig. 4A.1. showing drusenoid PED (arrows).

epithelial detachments (drusenoid PEDs), as seen in Fig. 4A.2 (color photograph) and Fig. 4A.3 (SS-OCT).

Age-related Macular Degeneration (Late)

The late or advanced form of AMD can be classified into the neovascular and non neovascular forms. Non neovascular late AMD is defined by the presence of geographic atrophy, while the late neovascular form is characterized by the presence of choroidal neovascularization (CNV).

Age-related Macular Degeneration (Late) — Non Neovascular

Geographic Atrophy

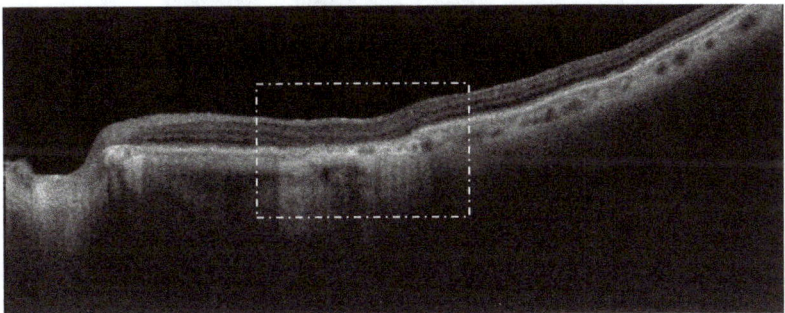

Fig. 4A.4.   SS-OCT of a patient with geographic atrophy.

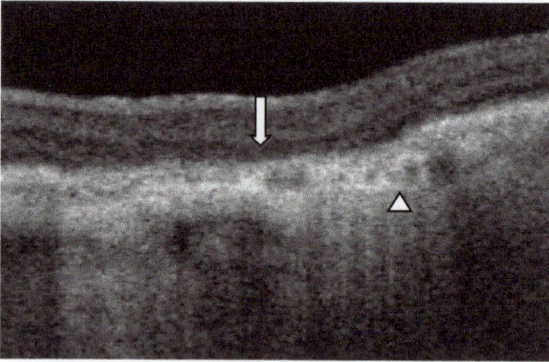

Fig. 4A.5.   Magnified view of the white box in Fig. 4A.4. Attenuation of the RPE line (arrowheads) with thinning of the inner retina and disruption of normal outer retina anatomy is seen (arrow). Thin choroid is a characteristic finding in patients with AMD.

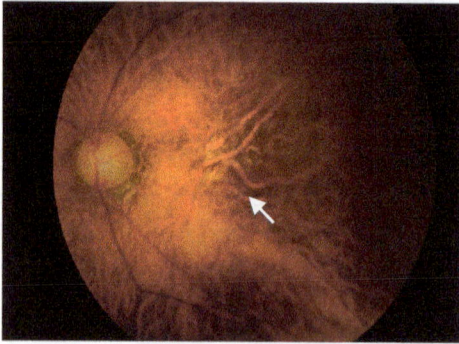

Fig. 4A.6.   Color fundus photograph showing geographic atrophy (arrow). Prominent choroidal vessels can be seen as a result of RPE atrophy.

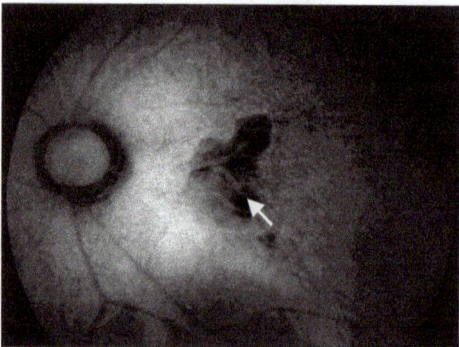

Fig. 4A.7.   Fundal autofluoresence reveals a hypoautofluorecent area (arrow) corresponding to the area of geographic atrophy.

OCT-A in Geographic Atrophy

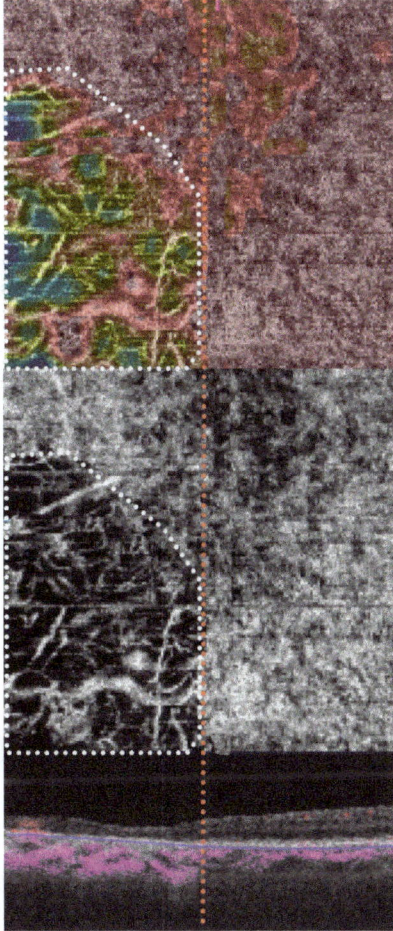

Fig. 4A.8.   OCT-A image of GA.
GA results in underlying hyper transmission as seen by increased flow-on angio B scan (bottom image). The unmasking artefact results in an area of highflow representing the underlying choroidal vessels seen in both the choriocapillaris layer on OCT-A (middle image) and on the density map (top image). The dotted red line denotes the border of atrophy from normal retina.

Age-related Macular Degeneration (Late) — Neovascular

Classification of AMD CNV

AMD CNV are abnormal blood vessels that originate from the choroid and are characterized by the layer in which they extend to. In Type I neovascularization, the CNV extends into the space between the Bruch's membrane and retinal pigment epithelium (RPE). In Type II neovascularization, the CNV extends into the subretinal space, and in Type III neovascularization or retinal angiomatous proliferation (RAP), the CNV extends from the choroid to form anastomosis with the deep retinal capillaries.[4]

Type I CNV

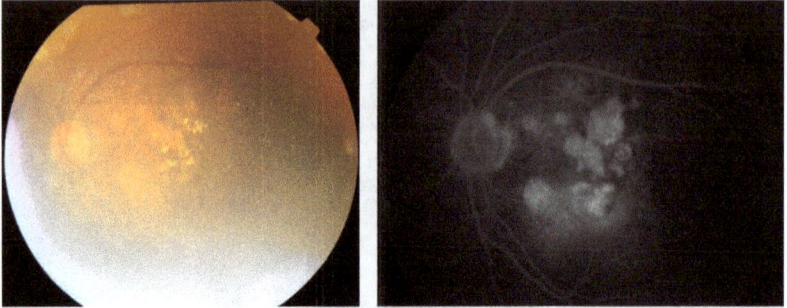

Fig. 4A.9. Color fundus photograph (left) of the Type I or occult CNV. The FFA image (right) shows late leakage from a fibrovascular PED (white).

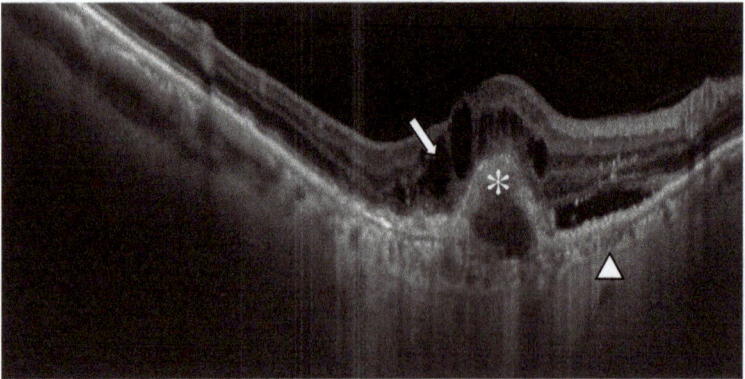

Fig. 4A.10. SS-OCT shows clear delineation of the RPE layer. The sub RPE hyper-reflective lesion (asterisk) represents a Type I (occult) CNV. Overlying retinal cysts (arrow) and adjacent subretinal fluid (arrowhead) are seen. FFA (Fig. 4A.9) shows a corresponding area of late leakage, confirming the diagnosis of occult CNV.

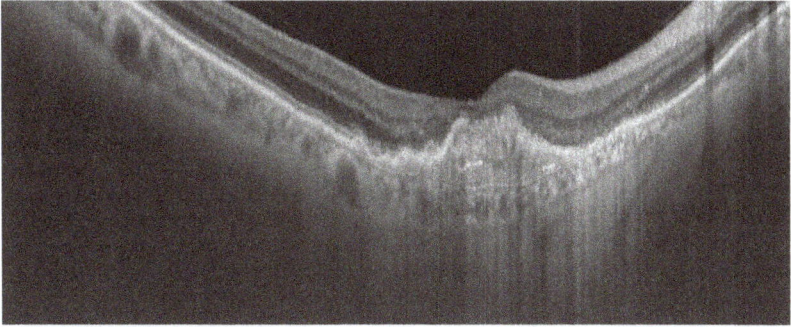

Fig. 4A.11. One month after treatment with intravitreal anti-VEGF, both intraretinal and subretinal fluid have resolved.

OCT-A in Type I CNV

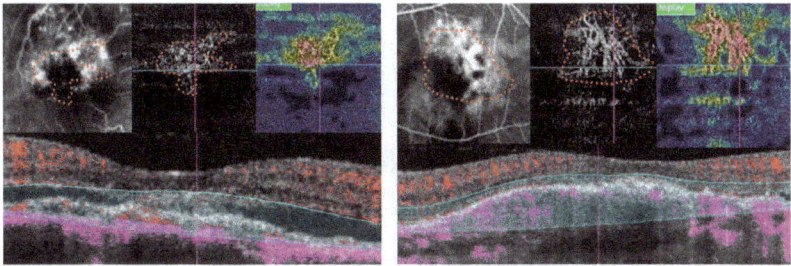

Fig. 4A.12. Type I CNV.
On FA (upper left image in both eyes), Type I CNV are when the abnormal vessels are confined under the RPE. The orange dotted line demarcates the extent of the CNV as seen on OCT-A. When superimposed on the FA images, the hyperflourscecent areas do not correlate well with the lesion as seen on OCT-A, indicating lesions which are either blocked by leakage or with ill-defined origins. The entire extent of the lesion, however, is well visualized on OCT-A despite being under the RPE. The lesions are more obviously seen on the density maps (color images right of the OCT-A scans). Warm colors represent areas of increased flow while cooler colors representing areas of reduced flow. Corresponding B scans (bottom images) with flow signals in red/purple show the level of the lesion below the RPE.

Type II CNV

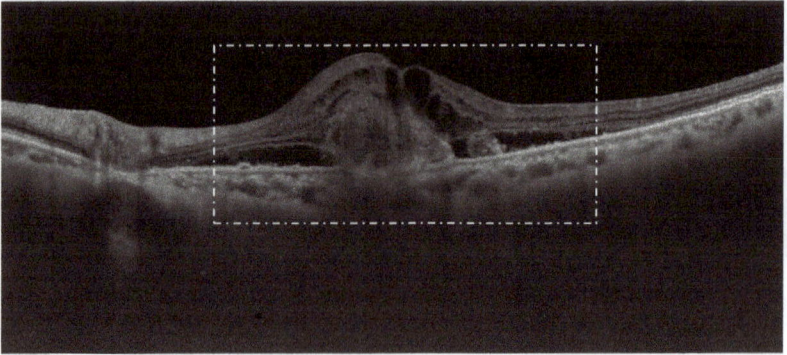

Fig. 4A.13. SS-OCT of a patient with Type II CNV.

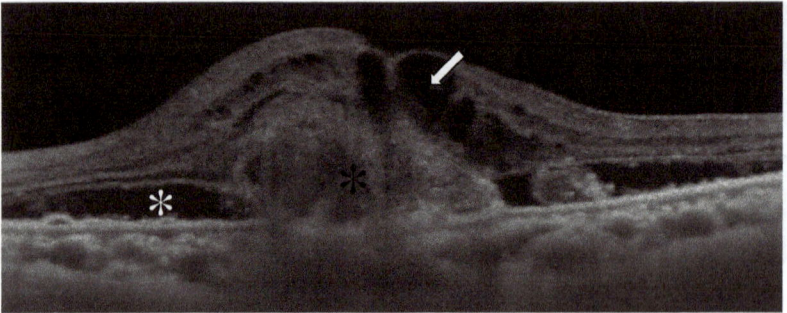

Fig. 4A.14. Magnified view of white box (Fig. 4A.13) shows the presence of SRF (white asterisk) and macula cyst (arrow) A subfoveal hyperreflective lesion localized within the subretinal space (black asterisk), representing a Type II CNV, is seen on SS-OCT. This corresponds to the area of early leakage on FFA (Fig. 4A.16), confirming the diagnosis of classic CNV.

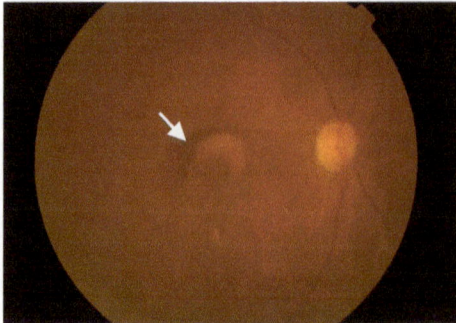

Fig. 4A.15. Corresponding color photograph to Fig. 4A.13 showing subretinal blood (white arrow).

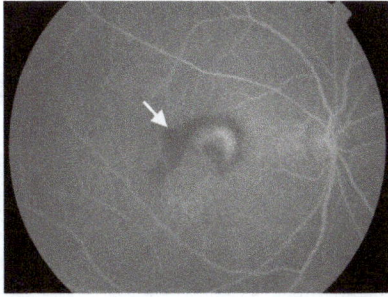

Fig. 4A.16. Corresponding FFA to Fig. 4A. 13 showing early phase hyperflourescent lesion (white arrow).

OCT-A in Type II CNV

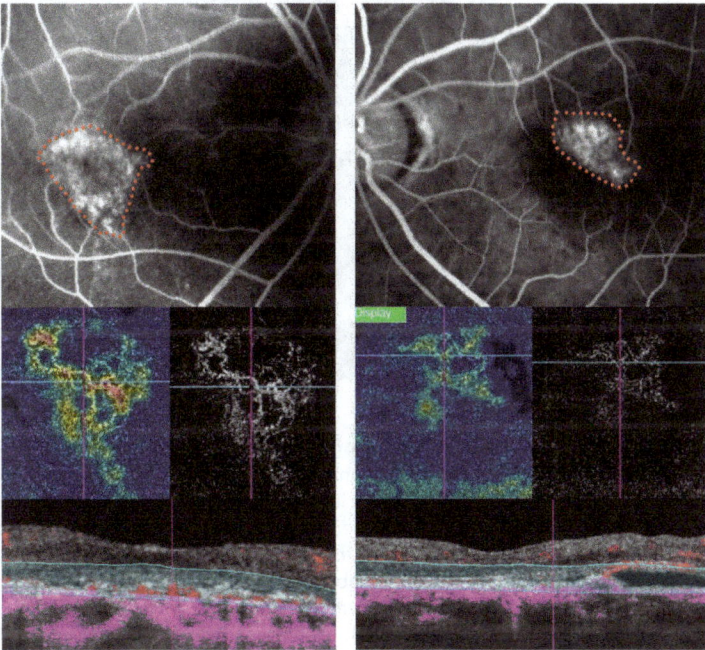

Fig. 4A.17. Type II CNV.
The Type II CNV occurs when the vessels proliferate in the subretinal space. This is well defined on fluorescein angiogram indicated by the orange dots (top images). These lesions correlate well with the OCT-A scans taken at the outer retinal slab (middle right image of each eye). The lesions are more obviously seen on the density maps (color images left of the OCT-A scans). Warm colors represent areas of increased flow while cooler colors represent areas of reduced flow. Corresponding B scans (bottom images) with flow signals in red show the level of the lesion above the RPE and in the subretinal space.

OCT-A in Type III CNV

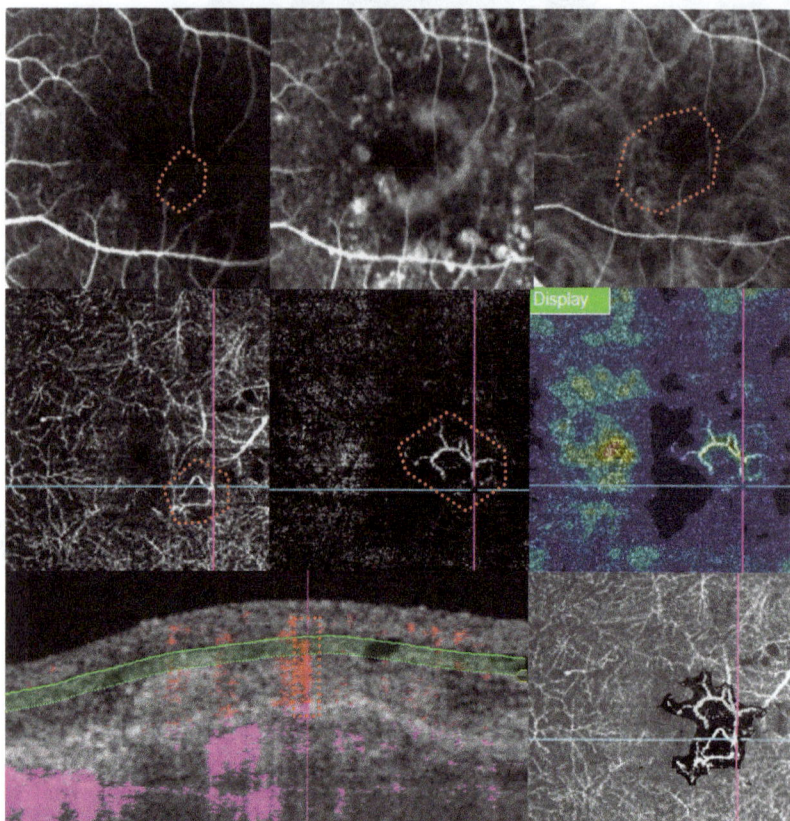

Fig. 4A.18.   Type III CNV.
The top left image is of the early frames of FA showing an abnormal vascular loop (orange dotted line). This appears to be connected to the choroidal circulation when correlated with ICGA (top right image). The late frame FA (middle top) shows diffused leakage which can be confused with a Type I lesion. The utility of OCT-A in imaging Type III CNV is particularly useful as seen in the series of images in the middle row. The middle left image shows the deep retina slab where the abnormal loop of vessel seen on FA is more obviously seen on OCT-A (outlined in orange dotted line). The middle image shows the outer retina slab showing the extension of the abnormal vessel which extends through the deep retina layer into the choroid. The density map is shown on the middle right image. The angio B scan (bottom left) shows the extent of the lesion from deep retina through to the RPE (orange dotted box). The image on the bottom right is a composite of the location and morphology of the abnormal vessel from the deep retina slab to the outer retina slab.

Progression and Response of AMD CNV

Structural cross section OCT has become the imaging modality of choice for monitoring of AMD CNV. This is due to its ease of use, speed of acquisition and information provided in terms of the activity of the CNV lesion. The decision for continuing anti-VEGF therapy and subsequent tempo of treatment is largely based on cross section OCT scans with the activity of the lesion denoted by the presence intra- or subretinal fluid.[5] Other clinical markers of activity include the presence of hemorrhage on clinical examination. Dye-based

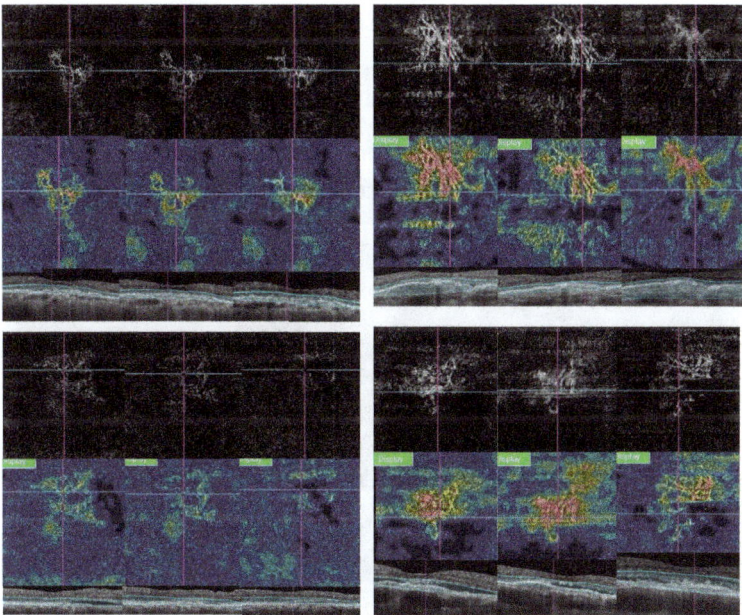

Fig. 4A.19. Progression of AMD CNV.
Four eyes showing different progression on OCT-A after anti-VEGF therapy. The series on the left show images of lesions that respond to anti-VEGF therapy. In the top left series of images, 6 × 6 mm OCT-A scans were performed to capture the entire lesion. The left and middle column shows the lesion at baseline (left) and after the induction phase of 3 anti-VEGF treatments (middle). The OCT-A scan on the right shows the lesion 10 months after commencement of treatment. The lesion on OCT-A has reduced in size over time; however, there is still a persistence of abnormal vasculature which may possibly represent a stable non-active lesion. The series of scans on the bottom left show a lesion that has progressively shrunk in size and almost completely resolved with increasing areas of hypo flow with corresponding resolution of fluid on B scan. The time frame and treatment regimen is similar in both cases. Vessel density scans are included in both cases.

angiography, however, is rarely employed for routine monitoring of treatment response. Due to the non-invasive and relatively speedy acquisition, OCT-A holds great promise as an adjunctive imaging modality for monitoring of the disease.

The series of images on the right represent persistent lesions despite continuing anti-VEGF therapy. The top series of images show a lesion that is persistent with a paucity of fluid on B scan. Similar to the top left series, this may represent a stable non-active lesion which has not changed in size or morphology over time and with treatment. In the bottom right series, the lesion is persistent in size and morphology with persistent subretinal fluid on B scan. This is a Type I CNV which may be more resistant to anti-VEGF therapy.

OCT-A in Non Exudative Neovascular AMD

The presence of exudation and/or hemorrhage is essential in the diagnosis of neovascular AMD. Recent studies, however, have shown

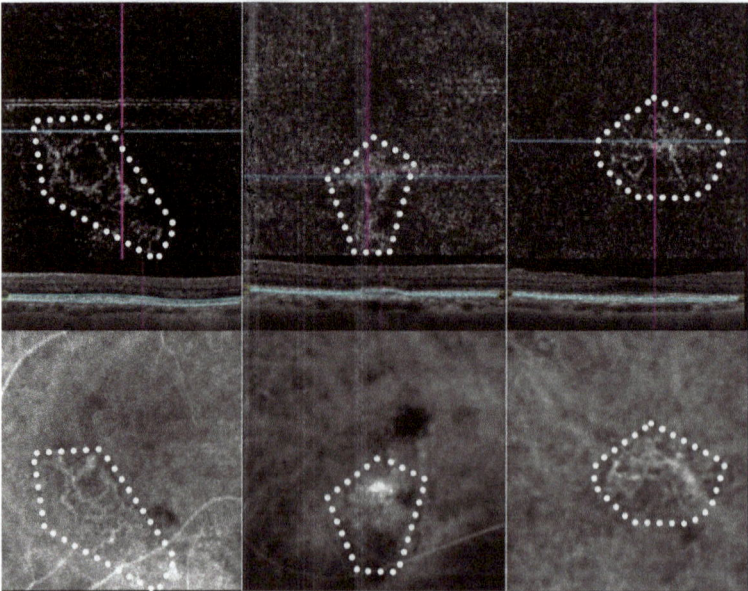

Fig. 4A.20.   Non exudative neovascular AMD.
Top row shows CNV lesions as imaged on OCT-A with corresponding B scans showing no signs of intra- or subretinal fluid. The bottom row shows ICGA imaging of the CNV lesions in the corresponding eyes.

the presence of "non-exudative neovascularization" in eyes with early or intermediate AMD.[6-8] In these reports, the detection of such lesions is with the use of ICGA; however, dye-based angiography for such patients in a routine clinical setting may not be practical. The non-invasive and rapidly acquired scans that OCT-A provides may prove to be an invaluable tool in the study of this phenotype.

References

1. Wong TY, *et al.*, (2008). The natural history and prognosis of neovascular age-related macular degeneration: A systematic review of the literature and meta-analysis. *Ophthalmology* **115**(1): 116–26.

2. Ambati J, *et al*, (2003). Age-related macular degeneration: Etiology, pathogenesis, and therapeutic strategies. *Surv Ophthalmol* **48**(3): 257–93.

3. Spaide RF, JM Klancnik Jr. and MJ Cooney (2015). Retinal vascular layers imaged by fluorescein angiography and optical coherence tomography angiography. *JAMA Ophthalmol* **133**(1): 45–50.

4. Gass JD (1994). Biomicroscopic and histopathologic considerations regarding the feasibility of surgical excision of subfoveal neovascular membranes. *Am J Ophthalmol* **118**(3): 285–98.

5. Regatieri CV, L Branchini and JS Duker (2011). The role of spectral-domain OCT in the diagnosis and management of neovascular age-related macular degeneration. *Ophthalmic Surg Lasers Imaging* **42 Suppl**: S56–66.

6. Roisman L, *et al.* (2016). Optical coherence tomography angiography of asymptomatic neovascularization in intermediate age-related macular degeneration. *Ophthalmology* **123**(6): 1309–19.

7. Capuano V, *et al.* (2017). Treatment-naive quiescent choroidal neovascularization in geographic atrophy secondary to nonexudative age-related macular degeneration. *Am J Ophthalmol* **182**: 45–55.

8. Palejwala NV, *et al.* (2015). Detection of nonexudative choroidal neovascularization in age-related macular degeneration with optical coherence tomography angiography. *Retina* **35**(11): 2204–11.

4B Polypoidal Choroidal Vasculopathy (PCV)

PCV is a subtype of AMD CNV characterized by polypoidal lesions arising from terminal dilatations with branching vascular networks (BVN) arising from the choroid. It has specific characteristic on clinical examination and on cross section OCT[1]; however, diagnosis is made on indocyanine green angiography (ICGA). This subtype is clinically important as previous studies have suggested poorer response to anti-VEGF monotherapy and the condition may benefit from combination therapy with photodynamic therapy (PDT). It also affects a younger age group as compared to typical nAMD and is more common in the Asian population.[2] The use of OCT-A imaging for PCV is shown to detect the BVN in the large majority of cases; however, it can only image 15–30% of the polypoidal structures. This is thought to be due to the unusual blood flow inside the polypoidal lesions.[3]

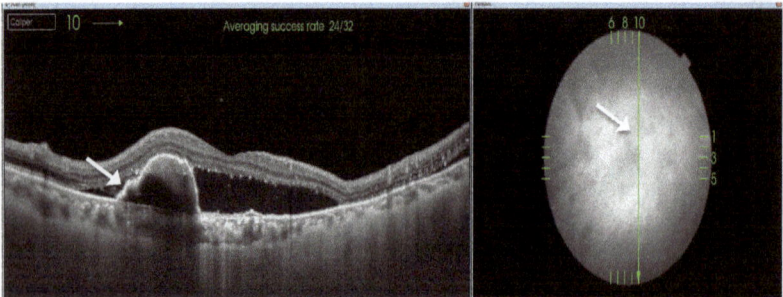

Fig. 4B.1. Swept source optical coherence tomography (SS-OCT) shows the presence of "double humps sign" (white arrow) associated with subretinal fluid.

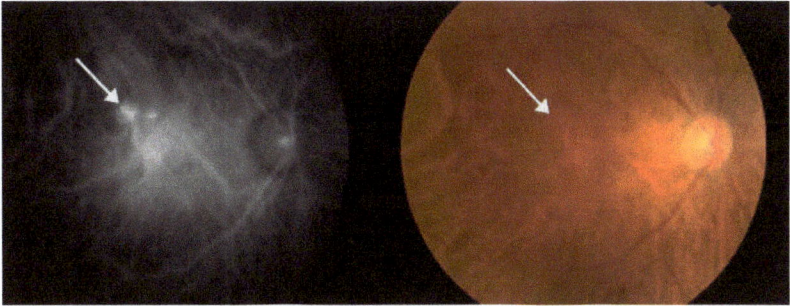

Fig. 4B.2. ICGA and color fundus photograph of PCV.
The presence of several hyperfluorescent nodules (white arrow) which persist into the late phase is seen on ICGA (left image) with corresponding presence of orange red nodule (white arrow) in color fundus photograph (right image).

Treatment Options for PCV

1. Photodynamic Therapy (PDT)
2. Anti-Vascular Endothelial Growth Factor (anti-VEGF)
   a. Bevacizumab (Avastin), Ranibuzumab (Lucentis), Aflibercept (Eylea)
3. Focal Argon Laser

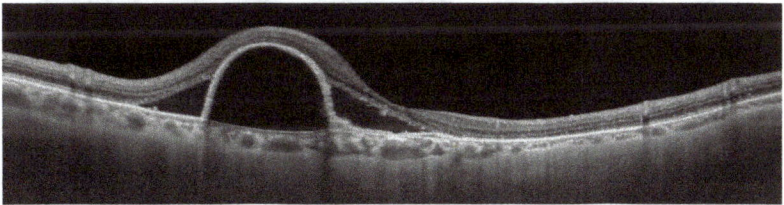

Fig. 4B.3. Decrease of subretinal fluid on post-treatment month 3.

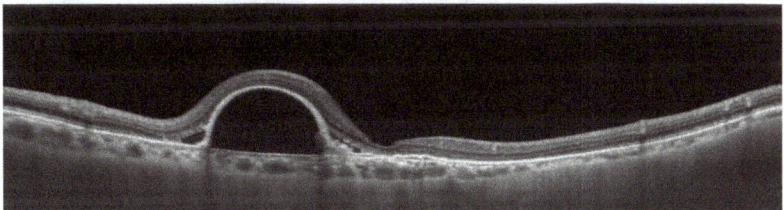

Fig. 4B.4. Significant improvement of subretinal fluid on post-treatment month 6.

OCT-A in PCV

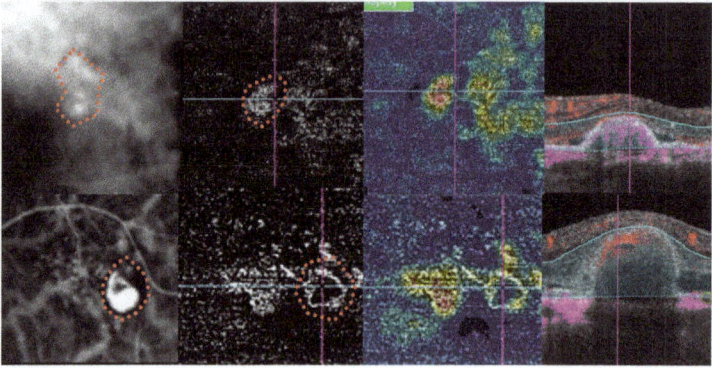

Fig. 4B.5.   OCT-A in PCV.
Two eyes (rows) with PCV imaged with ICGA and OCT-A. The top left image shows ICGA of a PCV lesion with associated BVN. The top left middle column shows the OCT-A in the outer retina slab. In this scan, some of the polyps and BVN can be visualized on OCT-A; however, the full extent of the polypoidal lesions do not show on OCT-A (denoted by the red dotted line on ICGA and corresponding red dotted line on OCT-A). The bottom row shows another eye where the BVN is well visualized on OCT-A; however, the large polypoidal lesion seen on ICGA shows as a hypoflow round structure (red dotted line). The corresponding density map scans are shown (right middle column). The left right column shows the corresponding angio B scans of both eyes.

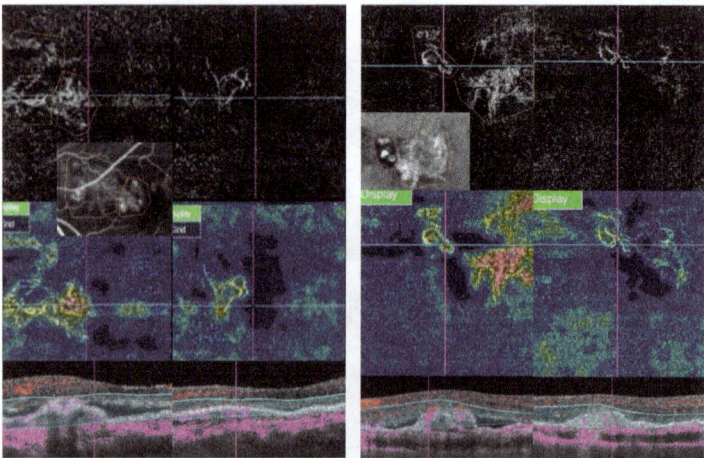

Fig. 4B.6.   Two eyes (left and right series of images) showing progression of PCV post combination treatment with anti-VEGF therapy and PDT.
Serial OCT-A scans of outer retina slab shows a reduction in size in both the BVN (orange dotted line) and polypoidal components (red dotted lines) of the lesion after 3, monthly anti-VEGF treatments and PDT. There is also corresponding reduction in subretinal fluid in the angio B scans.

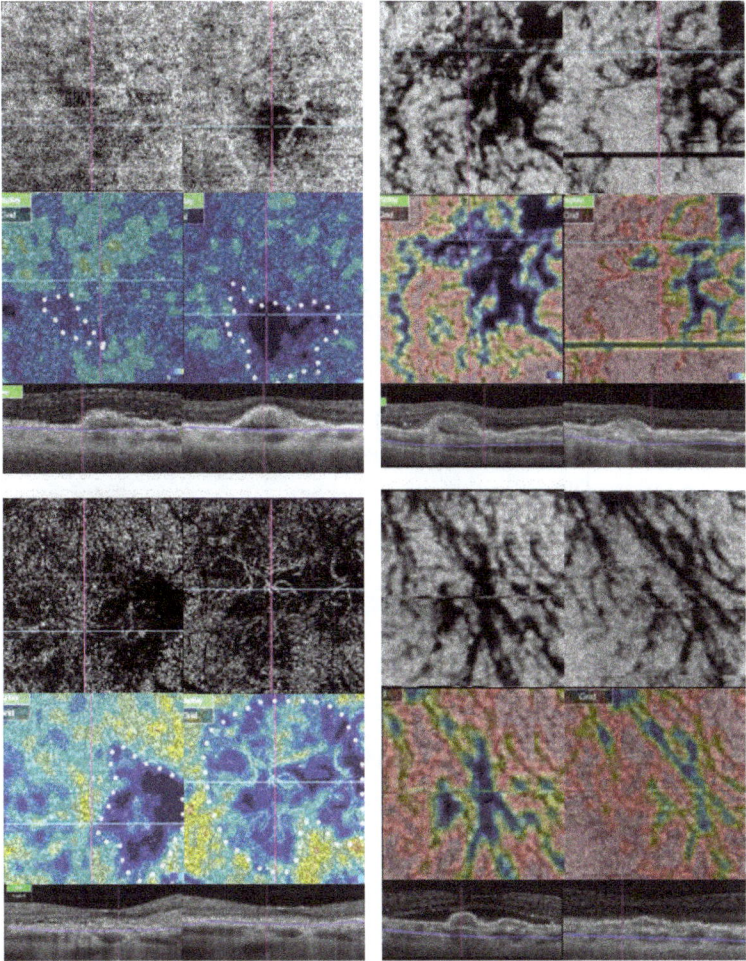

Fig. 4B.7. OCT-A and the choroidal layers in PCV.
The utility of the swept source OCT-A is shown when assessing the deeper layers of the retina and choroid especially in cases of PCV treated with different modalities. The series of images on the left show the alterations in the choriocapillaris layer on OCT-A after PDT and anti-VEGF combination therapy in 2 eyes. There is an enlargement area of reduced choriocapillaris density in the region of treatment, which is noted 3 months after PDT and 3, monthly anti-VEGF treatments. These changes are not noted when treated with anti-VEGF monotherapy alone. The series of images on the right show the alterations in the Haller's layer on OCT-A after PDT and anti-VEGF combination therapy in 2 eyes. The Haller's layer is not a preset automated layer but is a user-defined layer which is 10.4 µm in thickness following the contour of the Bruch's membrane through Haller's layer pachy-vasculature. There is a significant reduction in the width of Haller's vessels post-PDT as seen on OCT-A, suggesting an effect on the deep choroidal vessels from PDT which are thought to contribute to the pathogenesis of PCV.[4]

References

1. Ting DSW, Cheung G, Lim L and Yeo I (2015). Comparison of swept source optical coherence tomography and spectral domain optical coherence tomography in polypoidal choroidal vasculopathy. *Clin Experiment Ophthalmol.* DOI: 10.1111/ceo.12580.
2. Wong CW, Wong TY and Cheung CMG (2015). Polypoidal choroidal vasculopathy in Asians. *J Clini Med* **4**(5): 782–821.
3. Wang M, *et al.* (2016). Evaluating polypoidal choroidal vasculopathy with optical coherence tomography angiography. *Invest Ophthalmol Vis Sci* **57**(9): 526–32.
4. Teo KYC, *et al.* (2017). Comparison of optical coherence tomography angiographic changes after anti-vascular endothelial growth factor therapy alone or in combination with photodynamic therapy in polypoidal choroidal vasculopathy. *Retina.*

5

Vitreomacular Interface Abnormalities

Epiretinal Membrane

An epiretinal membrane (ERM) is a cellular proliferation on the inner retinal surface (Fig. 5.1). The symptoms of ERM may range from benign asymptomatic to debilitating metamorhopsia and central visual loss. The majority are idiopathic in nature, whereas some may have secondary causes such as retinal tear/retinal detachment, retinal vascular diseases, for example, diabetic retinopathy and vascular occlusions, inflammation, trauma, previous retinal lasers, retinitis pigmentosa, etc.

According to Gass, ERM can be classified into Grade 0, Grade 1 and Grade 2.[1]

- Grade 0 (a.k.a cellophane maculopathy) — a translucent membrane with no underlying retinal distortion
- Grade 1 (a.k.a wrinkled cellophane maculopathy) — ERM with irregular wrinkling of the inner retina
- Grade 2 (a.k.a macular pucker) — an opaque membrane causing obscuration of underlying vessels and marked full-thickness retinal distortion.

Treatment

- Conservative — benign, asymptomatic patients
- Surgery — pars plana vitrectomy, epiretinal membrane and inner limiting membrane peeling +/– membrane blue

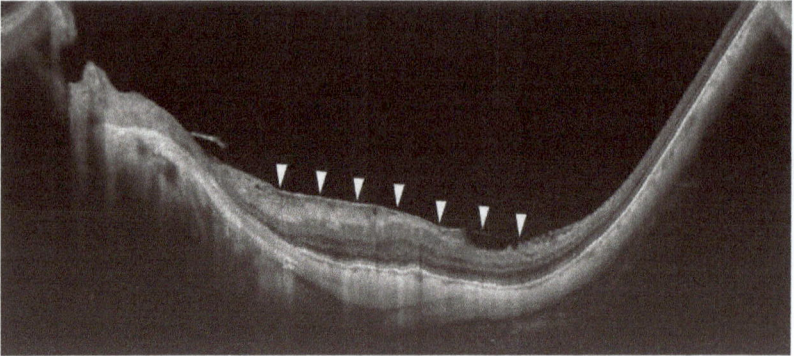

Fig. 5.1. SS-OCT shows epiretinal membrane (white arrows) in a high myope patient with posterior staphyloma.

Macular Hole

A macular hole (MH) is a retinal break involving the fovea. It is usually idiopathic but may be associated with myopia, epiretinal membrane and trauma. It can be caused by tangential traction and anterior-posterior traction of the posterior hyaloids on the parafovea. Up to 30% of patients have bilateral macular hole. The patients will complain of metamorphopsia or central scotoma and have positive Watzke-Allen's test (perception of a "break" in the middle of the slit beam).

According to Gass *et al.*, macula hole can be divided into four stages.[2]

- Stage 1a: foveal detachment (a.k.a yellow dot stage)
- Stage 1b: yellow ring
- Stage 2: full thickness macula hole (<400 μm)
- Stage 3: full thickness macula hole (>400 μm) with partial VMT
- Stage 4: full thickness macula hole (>400 μm) with posterior vitreous detachment

Treatment

- Conservative
- Surgical — can be considered for stage 2 or higher:
  Pars plana vitrectomy with epiretinal and inner limiting membrane peeling + endotamponade +/− triamcinolone, membrane or trypan blue or indocyanine green (ICG)

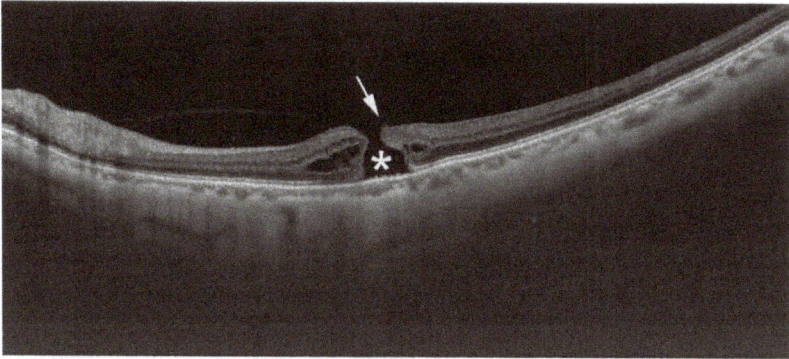

Fig. 5.2. SS-OCT shows presence of a full thickness macula hole (*) with vitreomacula traction (white arrow).

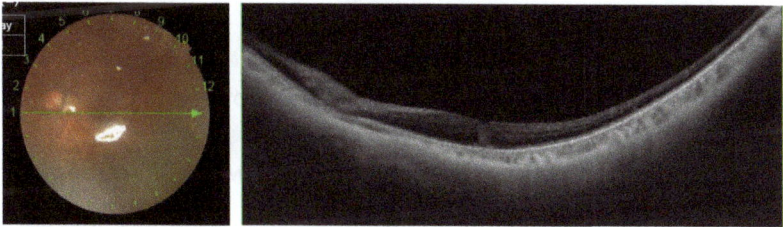

Fig. 5.3. A patient with full thickness macular hole of the left eye underwent ans pars plana vitrectomy, internal limiting membrane peeling and C_3F_8 gas tamponade. SS-OCT is able to demonstrate closure of macular hole on postoperative day 1 through a gas-filled eye.

Vitreomacular Traction (VMT)

Vitreomacular traction (Fig. 5.3) is a vitreoretinal interface disorder characterized by an incomplete posterior vitreous detachment with the persistently adherent vitreous over the macula, resulting in morphologic alterations and consequent decline of visual function.[3] With age, there is a sequential weakening of attachments between the vitreous and internal limiting membrane, starting from the perifoveal region, superior and inferior vascular arcades, fovea, mid-peripheral retina then optic disc.[4]

Based on the diameter of the vitreous attachment to the macular surface measured by OCT, the International Vitrecomacular Traction Study Group subclassified VMT into focal (1500 µm or less) or broad (1500 µm or more).[5]

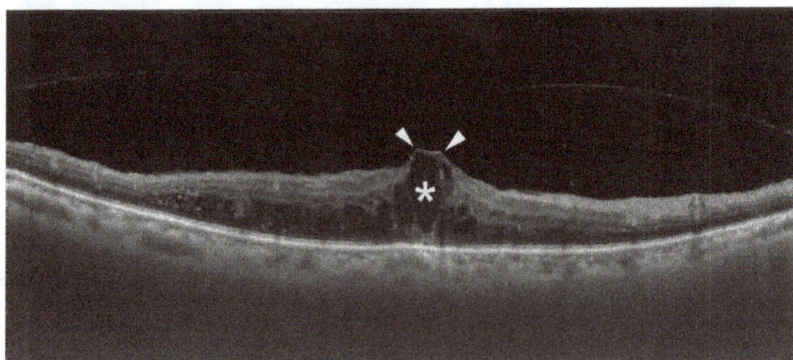

Fig. 5.4.   SS-OCT shows vitreomacular traction (VMT) denoted by 2 white arrows and cystoid macular edema (*).

Treatment

- Conservative — asymptomatic or mildly symptomatic patient with good visual acuity
- Surgery — pars plana vitrectomy combined with ERM peeling +/- ILM peeling and gas endotamponade
- New agent — ocriplasmin[6]

References

1. JDM G ed. (1987). Stereoscopic Atlas of Macular Diseases: Diagnosis and Treatment, 3rd ed. Mosby Year Book.
2. Johnson RN, Gass JD. (1988). Idiopathic macular holes. Observations, stages of formation, and implications for surgical intervention. *Ophthalmology* **95**(7): 917–24.
3. Reese AB, Jones IS, Cooper WC. (1970). Vitreomacular traction syndrome confirmed histologically. *Am J Ophthalmol* **69**(6): 975–7.
4. Johnson MW. (2010). Posterior vitreous detachment: Evolution and complications of its early stages. *Am J Ophthalmol* **149**(3): 371–82 e1.
5. Duker JS, Kaiser PK, Binder S, *et al.* (2013). The International Vitreomacular Traction Study Group classification of vitreomacular adhesion, traction, and macular hole. *Ophthalmology* **120**(12): 2611–9.
6. Stalmans P, Benz MS, Gandorfer A, *et al.* (2012). Enzymatic vitreolysis with ocriplasmin for vitreomacular traction and macular holes. *NEJM* **367**(7): 606–15.

6 Myopia

Pathologic myopia is defined as a myopic refractive error of more than 6 dioptres or axial length >26.5 mm with associated degenerative changes of the fundus.

Common degenerative changes include a tessellated fundus, chorioretinal atrophy (Fig. 6.2) lacquer cracks (Fig. 6.4), and posterior staphyloma. Factors such as vitreomacular traction (Figs. 6.1 and 6.6), posterior staphyloma and scleral stretching are thought to contribute to posterior pole abnormalities such as myopic foveoschisis (Fig. 6.6), fovea detachment (Fig. 6.7) choroidal neovascularization (Figs. 6.9 and 6.10), and dome-shaped macula (Fig. 6.5). The increased depth of penetration of the SS-OCT has enabled visualization of the posterior scleral border (Fig. 6.8), while the enhanced vitreous visualization (EVV) mode allows visualization of the posterior hyaloid and its attachments to the retina (Fig. 6.9).

SS-OCT provides an accurate high definition transverse image of these retinal changes allowing for prompt diagnosis and extent of the condition. Compared with SD OCT, SS-OCT has greater sensitivity and lower signal-to-noise ratio (SNR) at greater scanning depths. In addition, the enface scan function of the SS-OCT can help localize and determine the lateral extent of these changes.

OCT-A can be a useful adjunct for (1). Diagnosis and monitoring of myopic choroidal neovascularization (Figs. 6.12 and 6.13), and (2). differentiating between lacquer crack hemorrhage and choroidal neovascularization (Fig. 6.14).

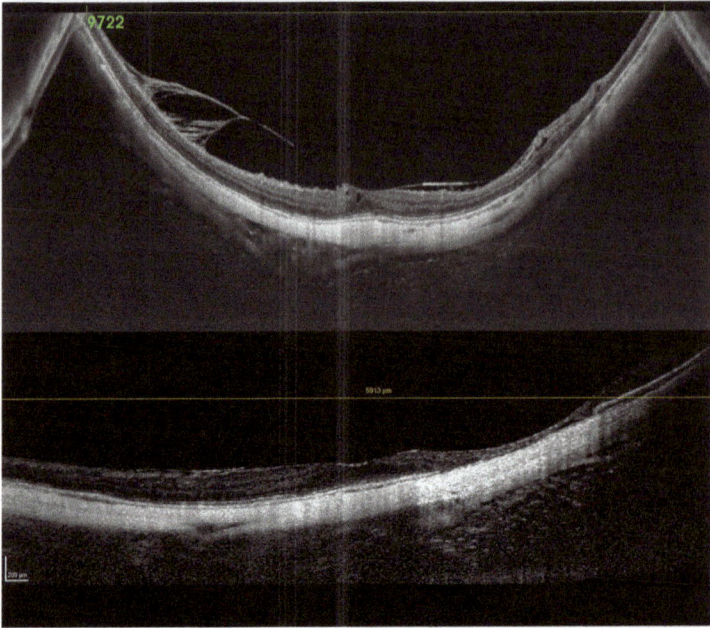

Fig. 6.1.   SS-OCT vs SD OCT.
Closest corresponding scans between SS-OCT and SD OCT are compared here. The useable area of the SS-OCT scan is 9.7 mm (top) as compared to the area imaged by SD OCT at 5.9 mm. In addition, the SS-OCT scan has greater sensitivity and lower signal-to-noise ration (SNR) than the SD OCT. Abnormal vitreoretinal adhesions are much more clearly visible on SS-OCT compared to SD OCT. Although a wider area is imaged in SS-OCT scans, mirror artifacts have been seen.

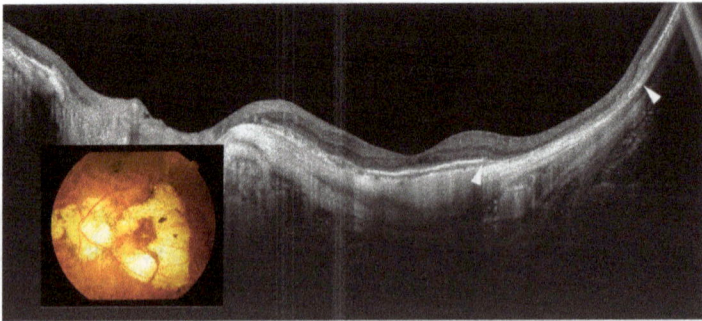

Fig. 6.2.   Chorioretinal atrophy.
Area of chorioretinal atrophy as seen in color fundus photography with corresponding SS-OCT scan (area between the 2 white arrows). SS-OCT scan shows decreased retinal thickness with an irregular ISOS juction (ellipsoid zone). This is in contrast to areas of normal retina beyond the 2 white arrows.

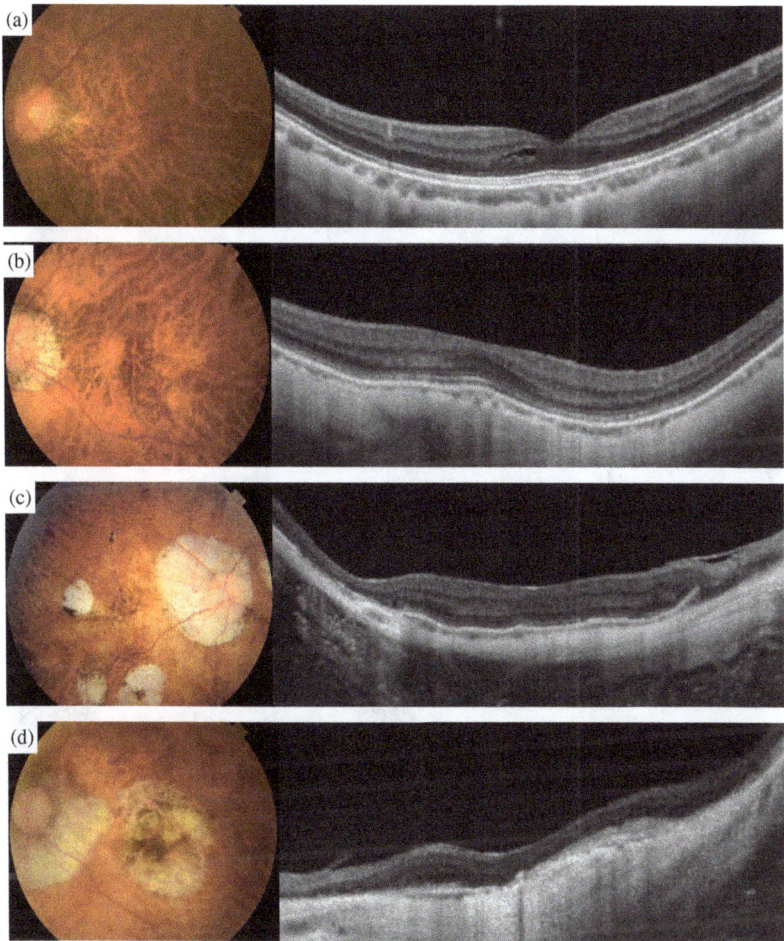

Fig. 6.3.   META-PM classification of myopic macular degeneration.
The META-PM classification is an international photographic classification and grading system for the severity of myopic macular degeneration. (a) META-PM category 1: tesellated fundus only, (b) META-PM category 2: diffuse atrophy, (c) META-PM category 3: patchy atrophy, and (d) META-PM category 4: macular atrophy. Note the gradual decrease in choroidal thickness with increasing severity of myopic macular degeneration.

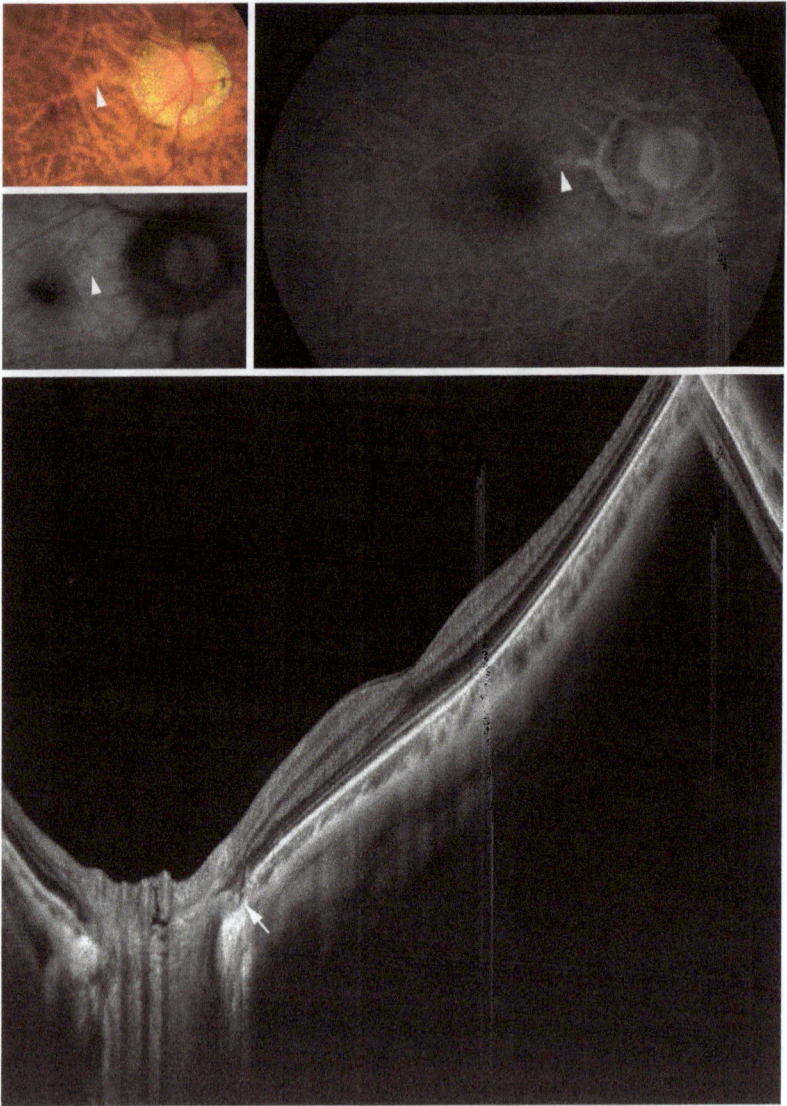

Fig. 6.4. Lacquer crack.
Top left images shows color fundus photograph and autofluorescence (AF) images of lacquer crack (white arrows) just temporal to the optic disc. Top right image shows late hyperfluorescence staining of lacquer crack on fluorescein angiogram. Corresponding SS-OCT scan shows disruption in the retinal pigment epithelium (bottom image, white arrow).

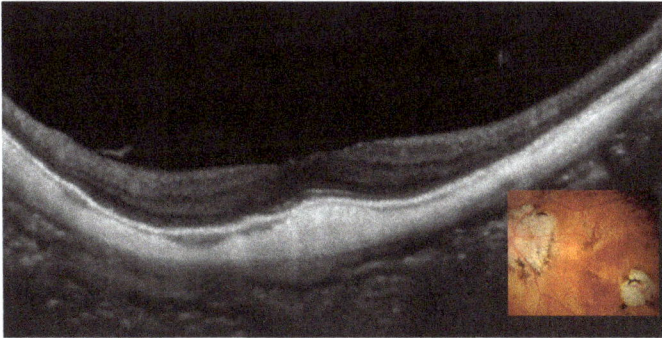

Fig. 6.5. Dome-shaped macula.
Longitudinal SS-OCT image of myopic fundus with dome-shaped macula. The subfoveal sclera is thickened with an anterior bulge, resulting in a dome-shaped appearance.

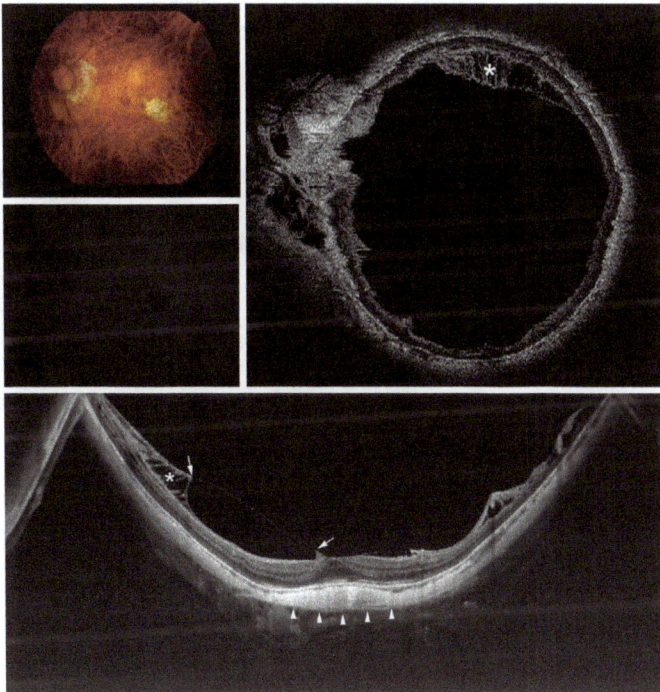

Fig. 6.6. Vitreomacular traction, retinoschisis, dome-shaped macula.
Top left images show color fundus photograph and AF images of eye with multiple myopic pathology. On SS-OCT longitudinal scan (bottom), VMT is clearly visualized (white arrows with tails), resulting in retinoschisis (denoted by *), showing strands of intraretinal bridges. Enface scan (top right) shows the intraretinal schisis as a honeycomb pattern (denoted by *).

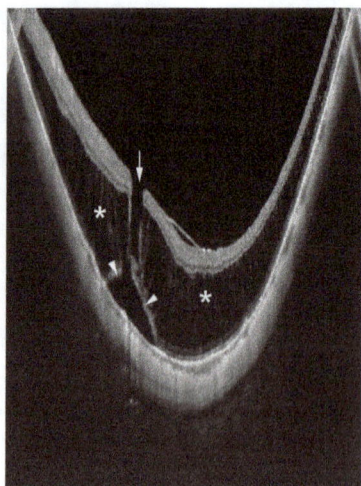

Fig. 6.7.   Foveoschisis with fovea detachment and macula hole.
This longitudinal SS-OCT scan shows foveoschisis (*) with fovea detachment (white arrows) and a macula hole (white arrow with tail).

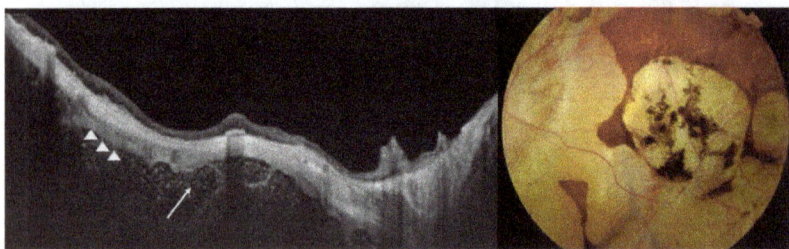

Fig. 6.8.   Visualization of the posterior scleral border.
In this eye with macular atrophy, the posterior scleral border is clearly visible on SS-OCT. The posterior scleral border is noted to be irregular in contour. The episclera (arrowheads) and even orbital fat (white arrow) can be seen on the SS-OCT scan.

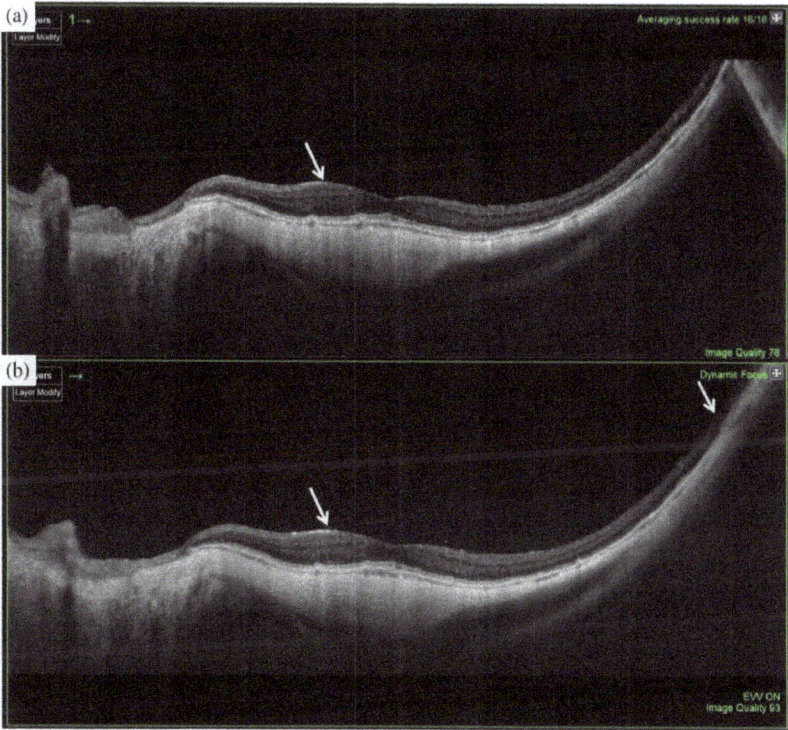

Fig. 6.9. Visualization of the posterior hyaloid.
(a) In this radial SS-OCT scan of an eye with posterior staphyloma and partial posterior vitreous detachment, the posterior hyaloid can be seen attached to the nasal perifoveal region (white arrow) but its temporal attachment is too anterior to be seen. (b) The dynamic focus mode with enhanced vitreous visualization (EVV) allows both nasal and temporal attachments (white arrows) to be seen.

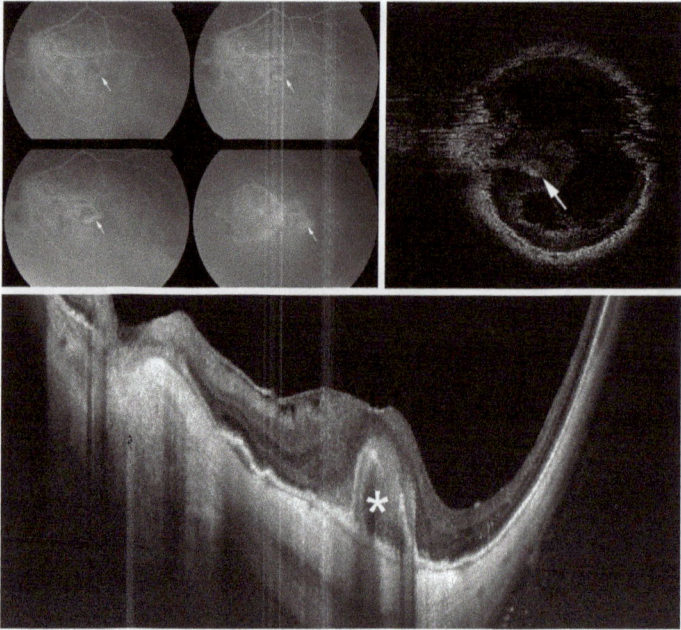

Fig. 6.10. Myopic CNV.
Top left images shows color fundus photograph and FFA of myopic CNV lesion (white arrows). This corresponds to the enface image on SS-OCT (top right) and longitudinal OCT scan (bottom). CNV lesion is denoted by (*).

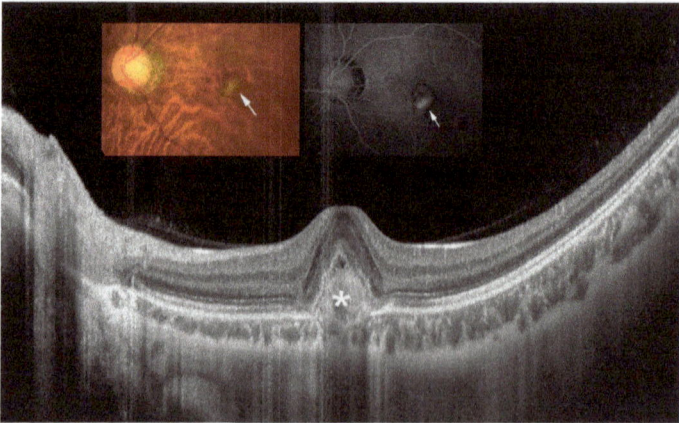

Fig. 6.11. Myopic CNV.
Inset images shows color fundus photograph and FFA of myopic CNV lesion. (white arrows). CNV lesion is denoted by (*) in longitudinal SS-OCT scan.

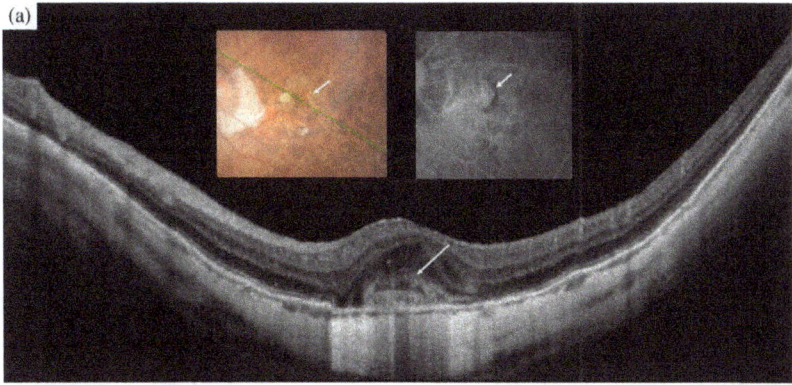

Fig. 6.12a. Myopic CNV.
A 62-year-old highly myopic Chinese male (–8.5 D) presented with sudden blurring of vision and metamorphopsia in his left eye. Inset images show color fundus photograph and FFA of myopic CNV lesion (white arrows). The longitudinal SS-OCT scan shows a subfoveal type 2 CNV lesion in the subretinal space.

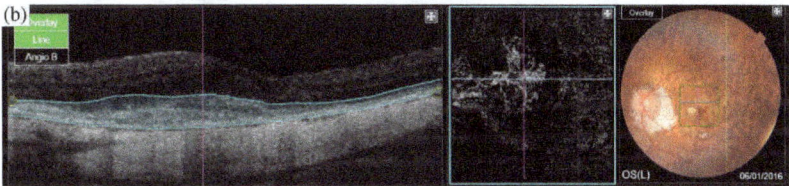

Fig. 6.12b. OCT-A appearance of myopic CNV.
3 × 3 mm OCT-A scan of the same patient above. The green box (right) overlay on the fundus image denotes the scanned area. Note the blue segmentation lines on the OCT-B scan (left), demarcating the outer retina slab and correctly segmenting the CNV lesion. Enface OCT-A image (middle) shows a net of tortuous lacy vessels within the outer retina slab representing the CNV membrane.

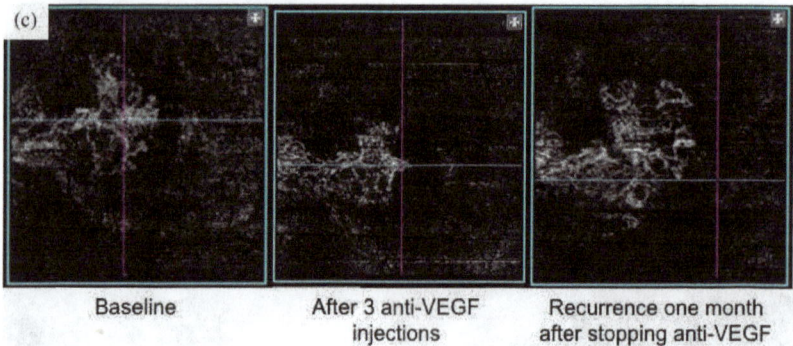

Baseline   After 3 anti-VEGF injections   Recurrence one month after stopping anti-VEGF

Fig. 6.12c.   Monitoring of CNV activity with OCT-A.
3 × 3 mm OCT-A scans of the same patient. The left panel shows the baseline appearance of the CNV membrane, decreasing in size and tortuosity noticeably after 3 anti-VEGF injections given over 3 months. One month after cessation of anti-VEGF treatment, CNV recurred and this can be observed on OCT-A as a significant increase in size of the vessel network (right panel).

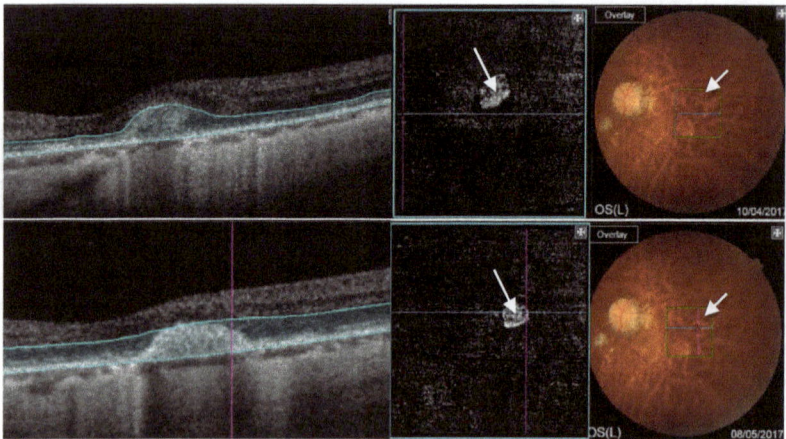

Fig. 6.13.   Monitoring CNV activity on OCT-A.
A 53-year-old myopic (–6.0 D) Chinese female presented with left eye metamorphopsia. Top left: OCT-B scan shows a subretinal hyperreflective lesion representing a type 2 CNV membrane. The blue lines indicate the segment from which the enface OCT-A scan is derived (top middle), in which the CNV membrane is observed as a tight net of hyperreflective flow signals (white arrow). Top left: the myopic CNV lesion (white arrow) can be seen on the fundus image. The green box represents the scan area. Bottom row: the same patient after treatment with an injection of anti-VEGF. Bottow left: OCT-B scan shows consolidation of the CNV lesion. Note that the margins of the lesion are now distinct, compared to the fuzzy margins pre-treatment. Bottom mid: the size of the CNV lesion on OCT-A has decreased but flow signals persist. Bottom left: fundus image now shows a pigmented scar (white arrow).

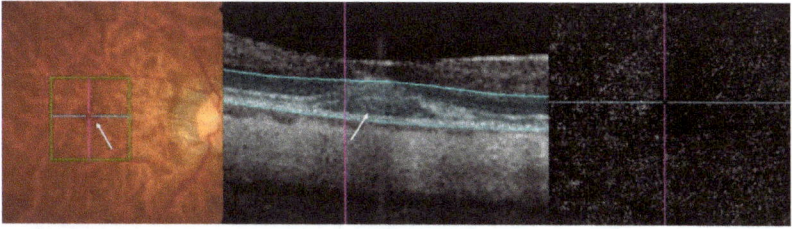

Fig. 6.14.   Lacquer crack hemorrhage.

OCT-A can be a useful adjunct for differentiating lacquer crack hemorrhage from myopic CNV. This 47-year-old myopic (–8 D) Chinese male presented with mild metamorphopsia in his right eye. A spot of hemorrhage can be seen on the fundus photograph (left panel, white arrow) lying across a faint linear yellowish streak. OCT-B scan (middle panel) shows a subfoveal hyperreflective lesion that extends from Bruch's membrane (white arrow) into the outer retina. OCT-A 3 × 3 mm scan (right panel) shows absence of flow signal corresponding to the lesion, indicating that the lesion was due to a lacquer crack hemorrhage.

7 Miscellaneous

Vogt-Koyanagi-Harada (VKH) Syndrome

VKH is a bilateral uveitis which frequently presents acutely as a bilateral posterior or pan-uveitis. It is associated with serous retinal detachments. In the chronic phase, the fundus shows a "sunset-glow" appearance. VKH is more common in pigmented races such as Asians, Hispanics and Native Americans. There is also a female predilection and occurs most frequently in the second to fifth decades of life. Intravenous corticosteroid in the early phase is the mainstay of treatment.

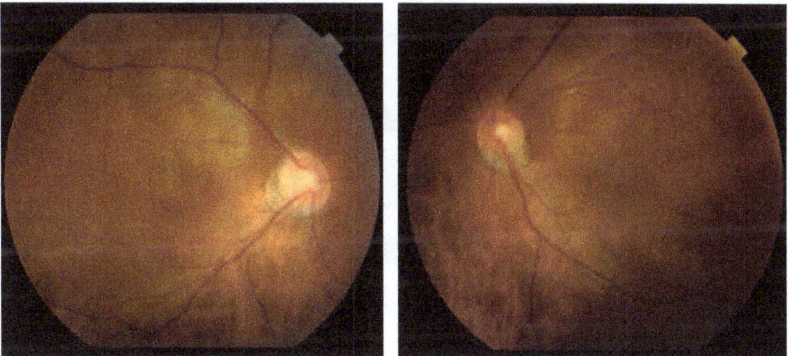

Fig. 7.1. Color fundus photographs showing acute VKH with multifocal serous retinal detachments.

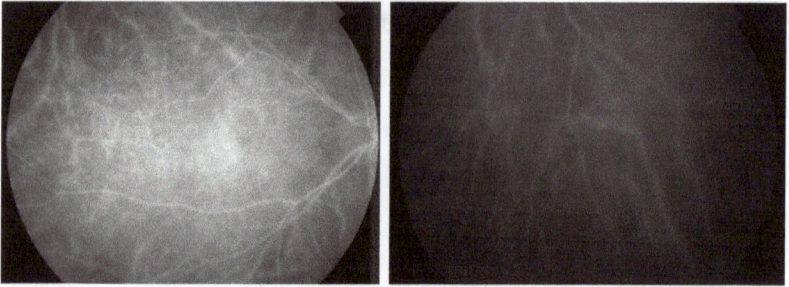

Fig. 7.2. Indocyanine green angiography showing delay of choriocapillaris perfusion, fuzzy choroidal vessels and perivascular leakage.

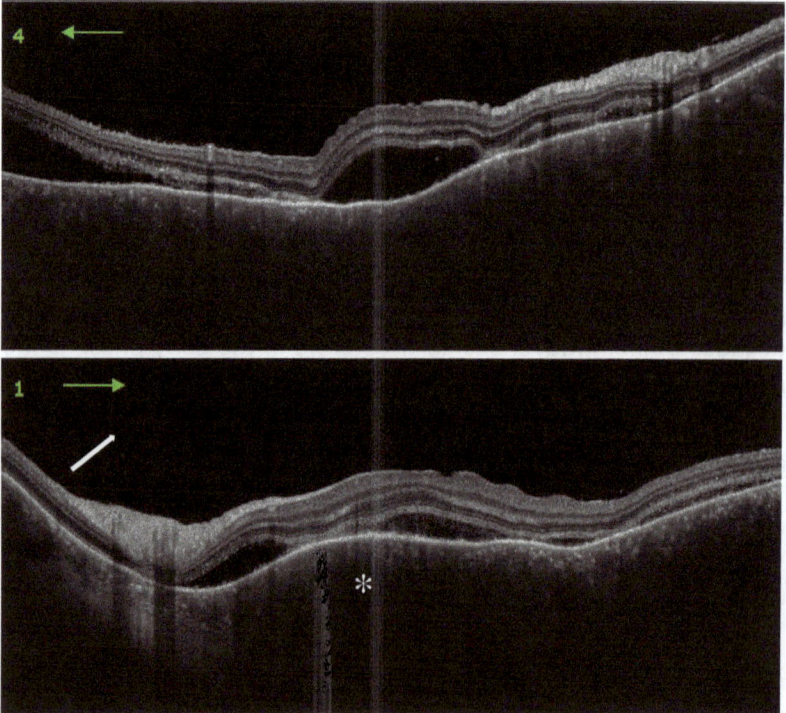

Fig. 7.3. SS-OCT not only shows presence of serous retinal detachments, but is also able to capture hyperreflective dots in the vitreous representing vitreous cells, as well as thickened choroid suggestive of choroidal inflammation.

Punctate Inner Choroidopathy (PIC)

PIC is an idiopathic inflammatory disorder of the choroid. It frequently affects young myopic females. Symptoms include loss of central visual acuity, photopsia and scotoma. Clinical findings include multiple small yellowish punctate lesions, without intraocular inflammation. PIC can be complicated by secondary choroidal neovascularization (CNV) and subretinal fibrosis. The visual prognosis is generally good and most patients can be observed. Treatment of inflammatory lesions close to fixation that has been reported include systemic, periocular or intraocular steroids or immunosuppressive agents and immunomodulators. If CNV develops, intravitreal anti-VEGF agents can be used.

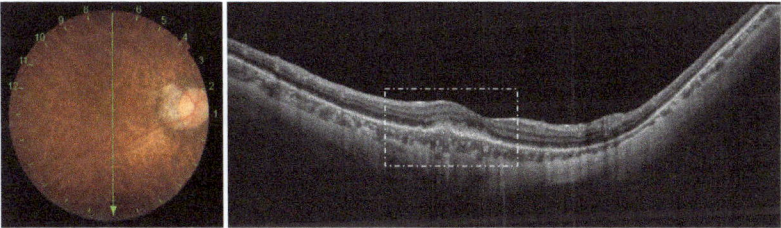

Fig. 7.4. A 40-year-old female presented with acute visual loss. Swept source OCT shows subfoveal active chorioretinal lesion with loss of ellipsoid zone, interdigitation zone and external limiting membrane.

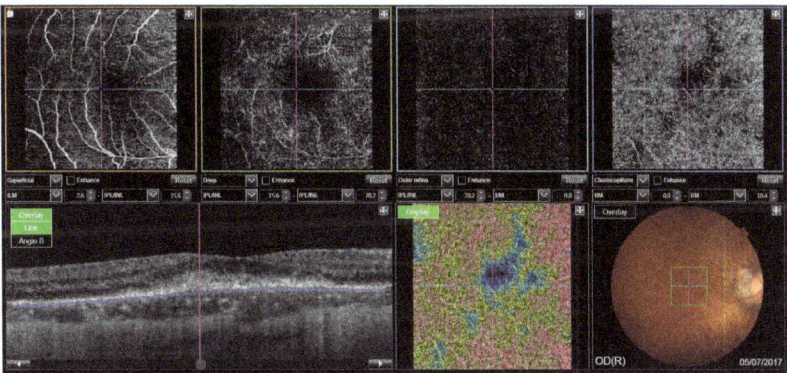

Fig. 7.5. 3 × 3 mm OCT-A of the same patient in Fig. 7.4 was useful to rule out the presence of secondary CNV.

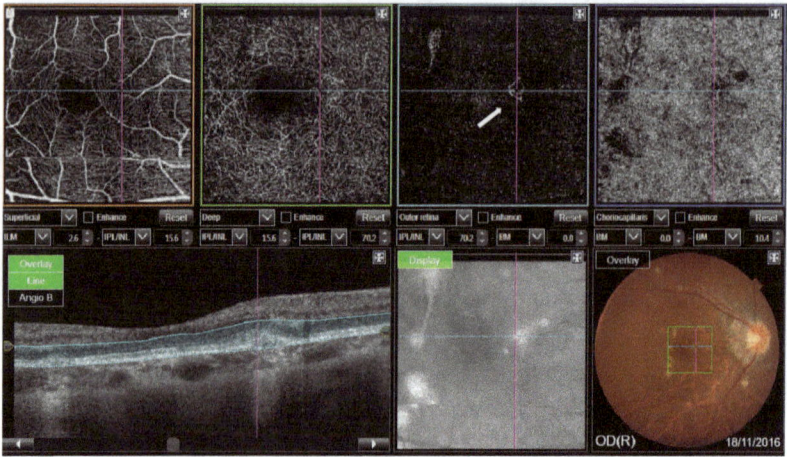

Fig. 7.6. A 35-year-old patient with PIC on the left eye. A secondary CNV can be seen (top row 2nd picture from right) in the outer retina layer (arrow). In the B scan image, an area of hyperreflectivity can be seen to corresponding to the CNV.

Choroidal Hemangioma

The choroidal hemangioma is a benign vascular tumor of the choroid. The circumscribed form is isolated and non-syndromic and the diffuse form is usually part of the Sturge-Weber syndrome. Visual symptoms are caused by exudative retinal detachment or degenerative changes in the macula. Treatment options include photodynamic therapy, transpupillary thermotherapy, radiation therapy or enucleation for painful blind eyes.

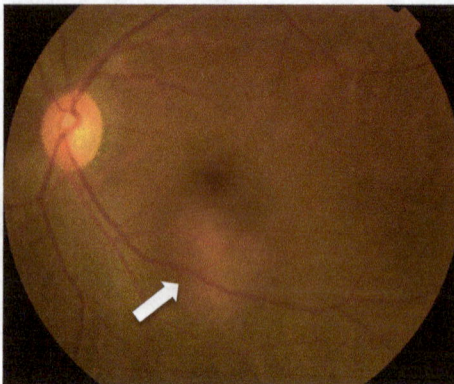

Fig. 7.7. A pale subretinal lesion which represents a circumscribed choroidal hemangioma is seen along the inferotemporal quadrant.

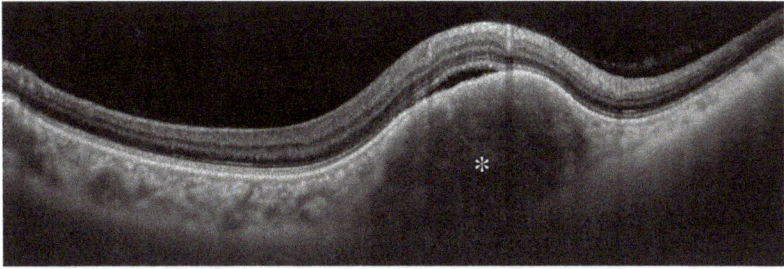

Fig. 7.8. SS-OCT shows a smooth, dome-shaped elevation of the choroidal tumor. There is also acoustic shadowing and expansion of the choriocapillaries. An overlying pocket of subretinal fluid is seen.

Choroidal Nevus

Choroidal nevi are the most common intraocular tumor seen in ophthalmic practice. It is a benign melanocytic tumor that is commonly seen after puberty. It is often asymptomatic and seen posterior to the equator. The clinical appearance is a brown or slate gray lesion that is flat or minimally elevated. It is occasionally associated with drusen or serous detachment of the sensory retina.

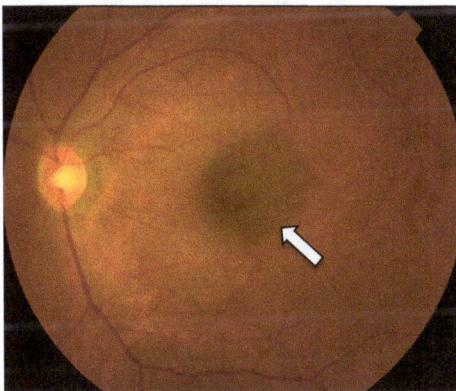

Fig. 7.9. Left eye choroidal nevus involving the macula.

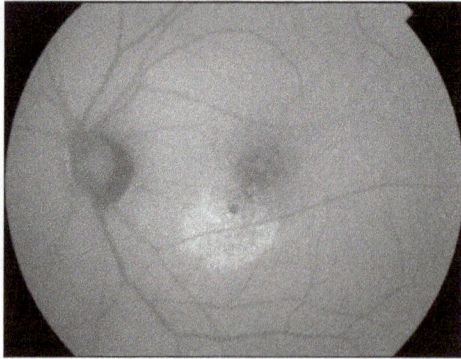

Fig. 7.10. Fundus autofluorescence showing hyperfluorescent areas suggestive of retinal pigment epithelium dysfunction.

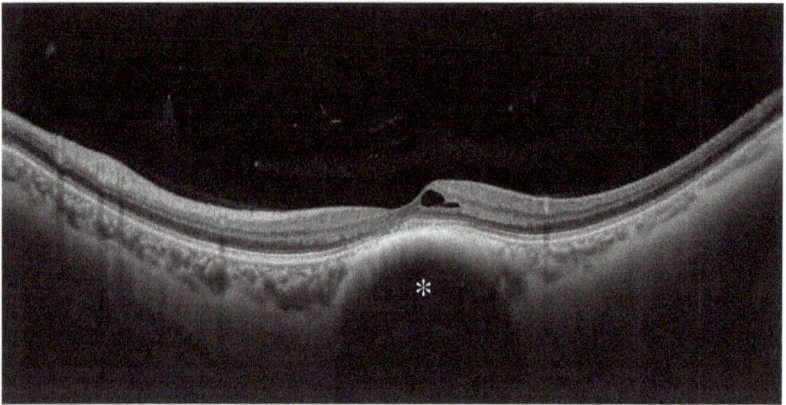

Fig. 7.11. SS-OCT image showing subfoveal choroidal nevus and overlying intraretinal cystoid changes. This image clearly shows the vitreous anatomy.

Choroidal Metastases

The choroid is the most common ocular site for metastases via hematogenous dissemination. The lung and breast are common sites for the primary tumor. Patients may present with blurred vision, photopsias or be entirely asymptomatic.

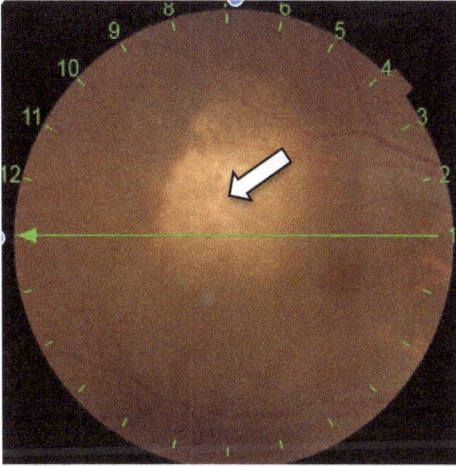

Fig. 7.12. This is a 49-year-old Chinese female with breast CA who complained of blurring of vision in her right eye. Fundus examination showed a yellowish subretinal mass (arrow) with ill-defined margins associated with subretinal fluid.

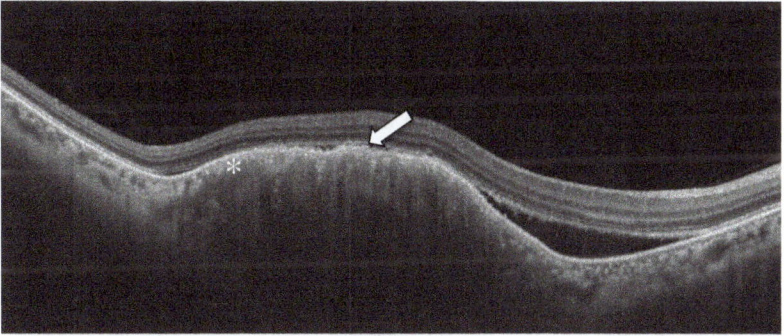

Fig. 7.13. SS-OCT image shows a "lumpy bumpy" subretinal mass with compression of the overlying choriocapillaris (asterisk), granularity of the ellipsoid zone (arrow) and subretinal fluid.

Arepalli S, Kaliki S, Shields CL. Choroidal metastases: Origin, features, and therapy. *Indian J Ophthalmol.* 2015;**63**(2):122–127. doi:10.4103/0301-4738.154380.

Focal Choroidal Excavation (FCE)

FCE is a newly described clinical idiopathic entity which manifests as thinning of the choroid. Vision is usually minimally affected. However, it can be associated with central serous chorioretinopathy, choroidal neovascularization, epiretinal membrane and age-related macular degeneration. OCT is the optimal imaging modality for diagnosis.

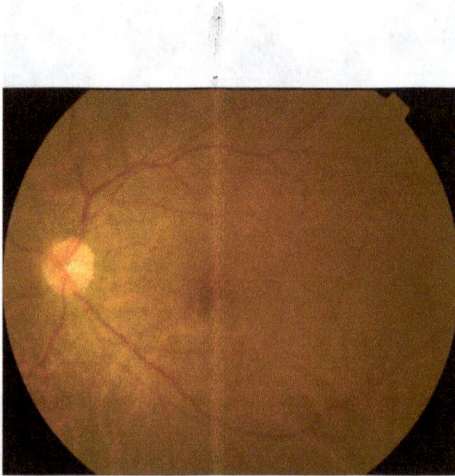

Fig. 7.14. Left eye showing mottling at the macula and retinal pigment epithelium disturbance.

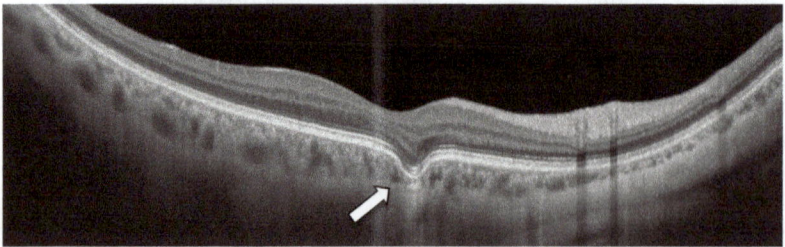

Fig. 7.15. SS-OCT of the left posterior pole showing thinned out choroidal tissue just beneath the area of FCE (arrow). This thinned out choroidal tissue has high internal reflectivity and there is poor visualization of both the medium- and large diameter choroidal vessels. There is loss of contour of the outer choroidal boundary, which appears to be pulled inwards by this abnormal choroidal tissue.

Morning Glory Disc

It is a congenital, funnel-shaped excavation of the posterior fundus that incorporates the optic disc. The disc is enlarged with a central glial tuft. The surrounding retinal vessels are anomalous and there is surrounding peripapillary chorioretinal pigmentary disturbance. It is often unilateral with a female preponderance. Ocular complications

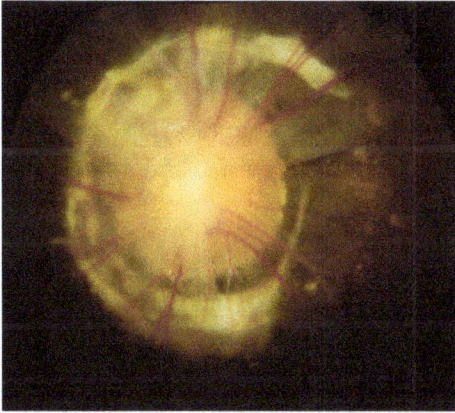

Fig. 7.16. Morning glory disc showing enlarged and excavated optic disc. The surrounding retinal vessels are in a spoke-like arrangement and there is an overlying tuft of glial tissue.

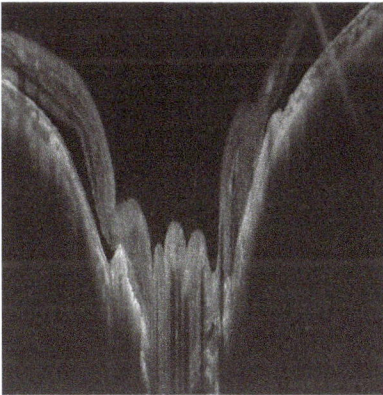 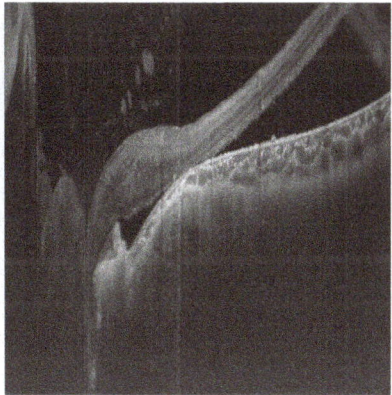

(a)   (b)

Fig. 7.17a and b. SS-OCT showing an excavated optic disc with surrounding retinal detachment.

include serous or rhegmatogenous retinal detachment, or choroidal neovascularization. Systemic associations include trans-sphenoidal basal encephalocele, corpus callosum agenesis and septo-optic dysplasia.

Choroidal Rupture

Choroidal rupture may occur in closed globe injuries where compression-decompression of the globe results in rupture of the inelastic Bruch's membrane and the underlying choroid. The damaged choriocapillaris and Bruch's membrane cause leakage of blood into the subretinal space. Choroidal ruptures may be associated with the development of choroidal neovascularization necessitating treatment with intravitreal anti-VEGF.

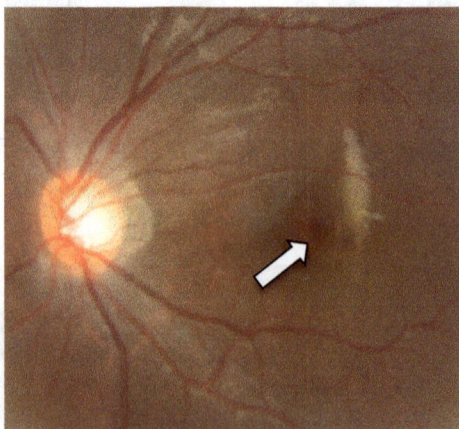

Fig. 7.18. 12-year-old boy who complained of sudden blurring of vision in his left eye after he was hit by a soccer ball. Fundus examination showed a yellowish white crescent-shaped streak just temporal to the fovea with subretinal hemorrhage (arrow).

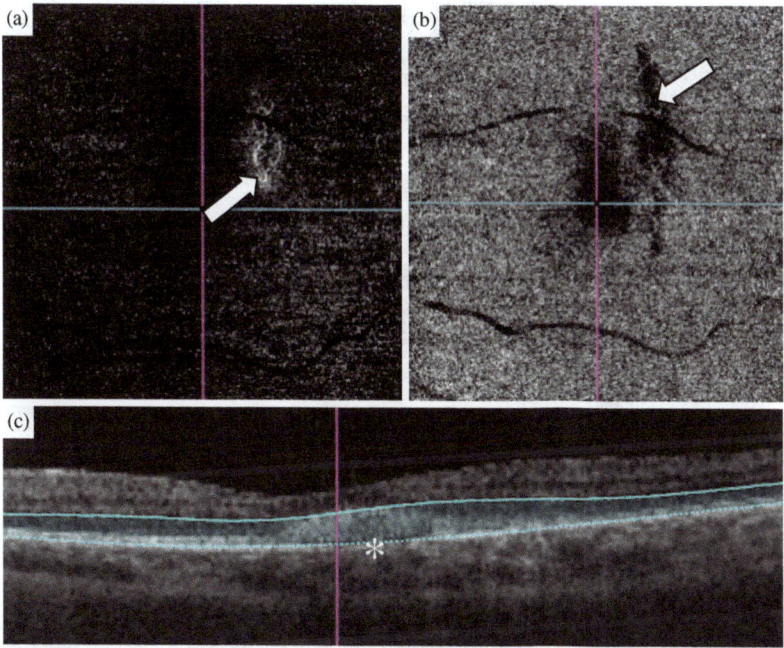

Fig. 7.19. OCT-A 3 × 3 mm scan of the same patient. Figure (a) shows the outer retina slab as represented by the blue segmentation lines on the OCT-B scan (c). Flow signals in this avascular segment (arrow) are suggestive of a choroidal neovascular membrane, visible also on the OCT-B scan as subretinal hyperreflective material (asterisk). On the choriocapillaris slab (b), the ruptured choroid is seen as a crescent-shaped area of signal void (arrow).

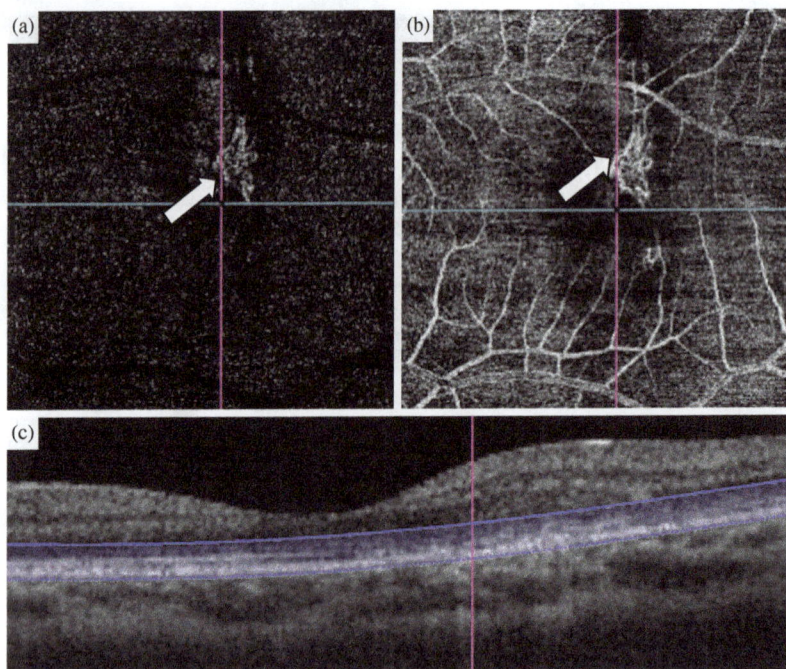

Fig. 7.20. OCT-A 3 × 3 mm scan of the same patient 1 month after intravitreal injection. Figure (a) shows the outer retina slab as represented by the blue segmentation lines on the OCT-B scan (c) and figure (b) shows the choriocapillaris slab. The subretinal hyperreflective material had resolved but the choroidal neovascular membrane is still visible as a tangle of flow signals (arrow).

8 Optic Nerve Conditions

Optical coherence tomography angiography (OCT-A) has recently gained growing interest in exploring the optic nerve microvasculature in various optic nerve conditions, such as primary open angle glaucoma (POAG), and in other neuro-ophthalmic conditions, including optic atrophy, as well as ischemic, inflammatory, hereditary optic neuropathies.[1] In POAG, OCT-A discloses localized attenuation of the optic nerve microvasculature, correlating with the disease severity and the associated visual field loss. In non-arteritic ischemic optic neuropathy, due to hypoperfusion in the capillary bed of the optic nerve head, OCT-A imaging may reveal segmental and global reduction of the peripapillary vascular flow density and peripapillary vascular tortuosity.[2] Systemic conditions, like multiple sclerosis, can be associated with reduced vascular flow explored with OCT-A, even in the absence of optic neurities. Further prospective studies are needed to determine the specificity of these OCT-A findings and whether they constitute the cause or the consequence of the underlying condition.

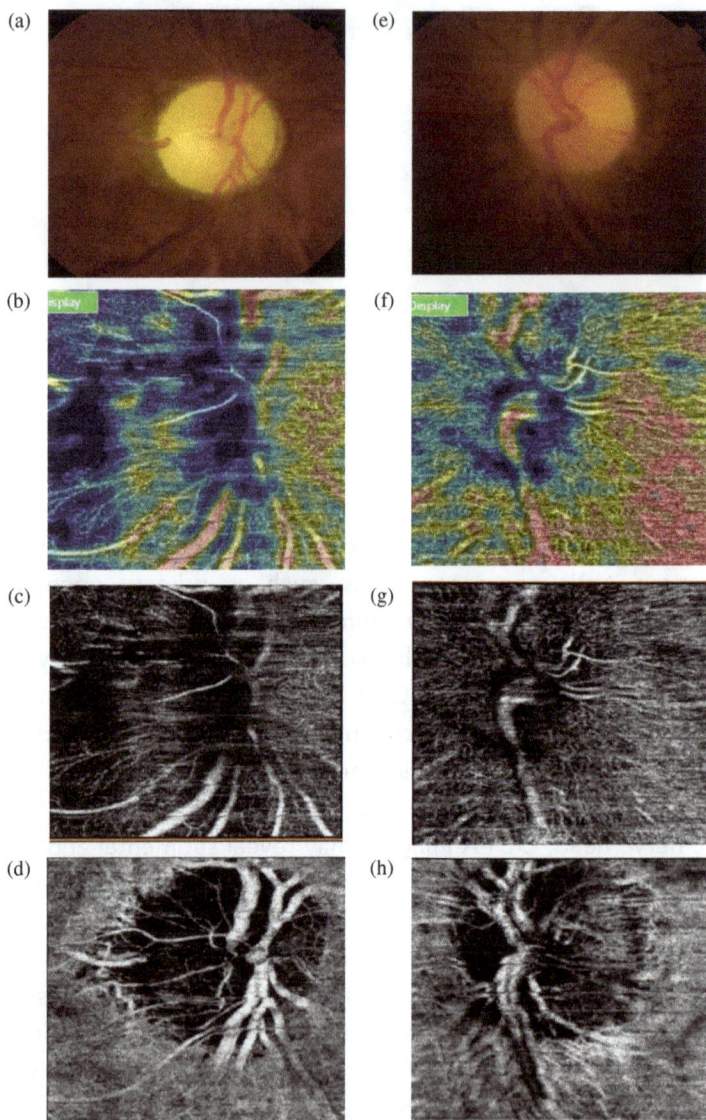

Fig. 8.1. Right optic atrophy (a), imaged with OCT-A (b-d), following the acute phase of anterior non ischemic optic neuropathy (NAION) in an elderly patient. The color fundus of the normal left optic nerve head (e) discloses a small cup/disc ratio, "at risk" for NAION. OCT angiography of the left optic nerve head shows a rich, dense, well distributed radial peripapillary capillary network (f-g). Palor of the right optic nerve head (a) is associated with decreased capillary density, predominantly on the temporal side, visible on color maps (d), and when exploring the superficial retinal capillary network (c). OCT-A at the choriocapillaris level shows reduction of the optic disc vascular network in the affected eye (d), compared to the healthy eye (h).

(a)

(b)

(c)

(d)

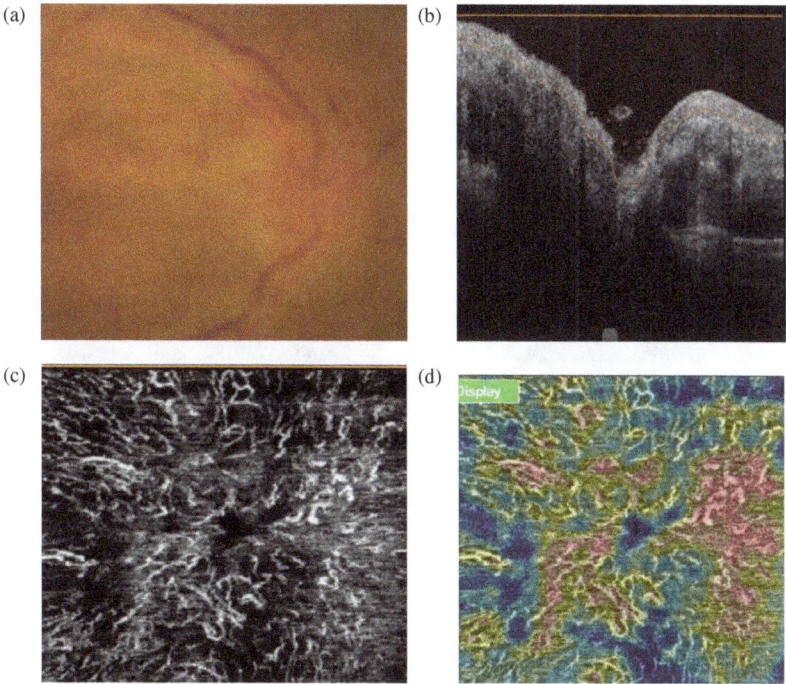

Fig. 8.2. Optic disc swelling and peripapillary hemorrhages (a) occuring in the right eye of a patient with infectious optic neuropathy due to tuberculosis. Elevation of the optic nerve head (b) is associated on OCT-A with superficial vascular tortuosity (c) and patches of less visible capillary network (in dark blue), on the color map (d). Decreased visibility of the vascular network (d) can be due to vascular hypoperfusion or as artifact due to optic disc edema

(a)

(c)

(b)

(d)

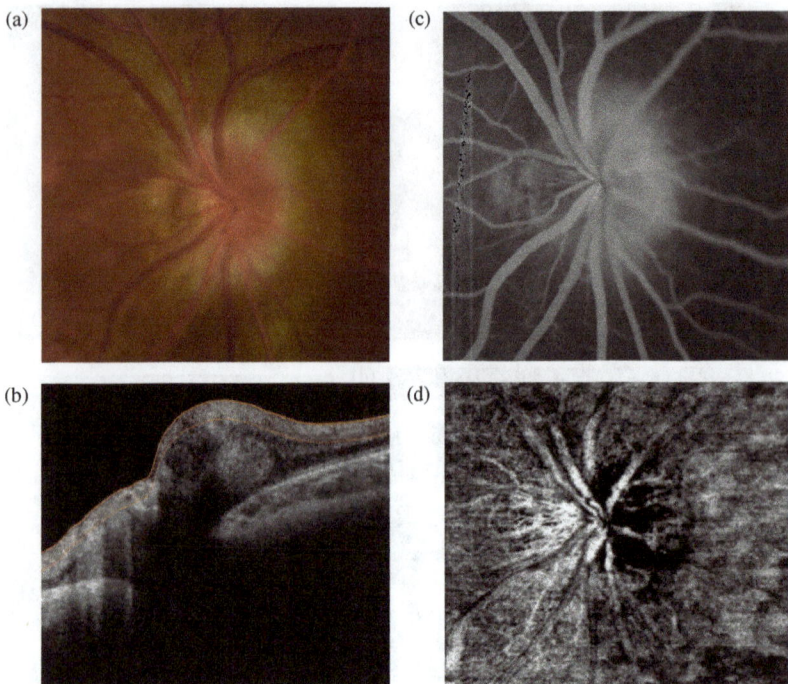

Fig. 8.3. Inflammatory oedematous right optic neuropathy. The right optic disc swelling (a), visible both on color fundoscopy and on a B-scan (b) is predominant in the nasal part of the disc, associated with localized diffusion on the late phases of the fluorescein angiogram (c). OCT-A of the right optic nerve head (d) discloses a correspondent area of less visible capillary network within the optic nerve head.

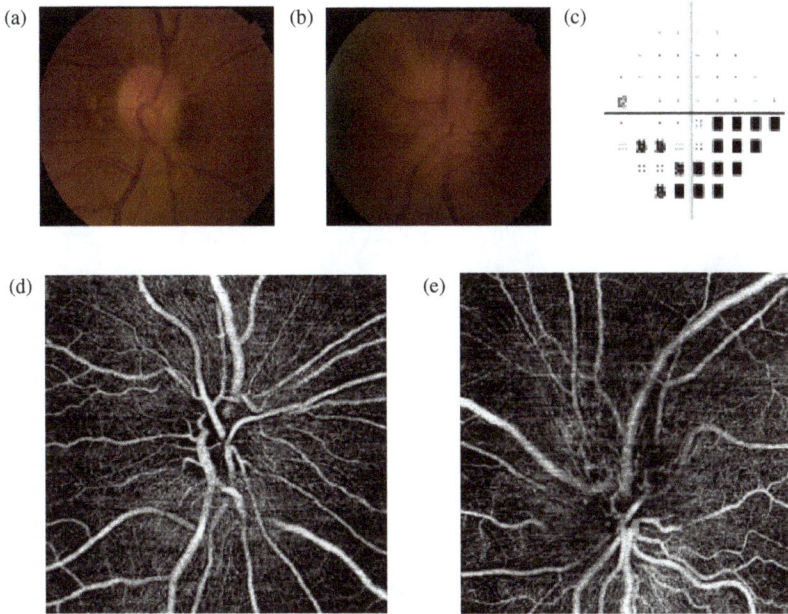

Fig 8.4. Acute non arteritic ischemic optic neuropathy (NAION), left eye, in a diabetic patient and other associated vascular risk factors. Color fundus pictures (a, b), displaying a crowded, disc at risk in the right healthy eye (a) and sectorial disc swelling in the affected left eye (b), causing an altitudinal inferior visual field loss (c). OCT-A displays a normal, dense, peripapillary capillary network in the right eye (d) and reduced flow density, predominantly in the upper area, in the affected left eye (e), corresponding to the inferior visual field defect.

References

1. Tan ACS, Tan GS, Denniston DK, Keane PA, Ang M, Milea D, Chakravarthy U, Cheung CMG (2018). An overview of the clinical applications of optical coherence tomography angiography. *Eye* **32**: 262–286.
2. Sharma S, Ang M, Najjar RP, Sng C, Cheung CY, Rukmini AV *et al.* (2017). Optical coherence tomography angiography in acute non-arteritic anterior ischaemic optic neuropathy. *Br J Ophthalmol* **101**(8): 1045–1051.

INDEX